SUSTAINABLE CITIES

Urban Planning Challenges and Policy

SUSTAINABLE CITIES
Urban Planning Challenges and Policy

Edited by
Kimberly Etingoff

AAP | APPLE ACADEMIC PRESS

Apple Academic Press Inc.
3333 Mistwell Crescent
Oakville, ON L6L 0A2
Canada

Apple Academic Press Inc.
9 Spinnaker Way
Waretown, NJ 08758
USA

Exclusive worldwide distribution by CRC Press, a member of Taylor & Francis Group

No claim to original U.S. Government works

Printed in the United States of America on acid-free paper

International Standard Book Number-13: 978-1-77188-318-4 (Hardcover)

International Standard Book Number-13: 978-1-77188-319-1 (eBook)

Library and Archives Canada Cataloguing in Publication

Sustainable cities (Apple Academic Press)
Sustainable cities : urban planning challenges and policy / edited by Kimberly Etingoff.

Includes bibliographical references and index.
Issued in print and electronic formats.
ISBN 978-1-77188-318-4 (hardcover).--ISBN978-1-77188-319-1 (pdf)
1. City planning--Environmental aspects. 2. Cities and towns--Environmental aspects. 3. Sustainable development. 4. Urban ecology (Sociology). I. Etingoff, Kim, editor II. Title.

HT166.S88 2016 307.1'216 C2015-907955-1 C2015-907956-X

Library of Congress Cataloging-in-Publication Data

Names: Etingoff, Kim, editor.
Title: Sustainable cities : urban planning challenges and policy / [edited by] Kimberly Etingoff.
Other titles: Sustainable cities (Apple Academic Press)
Description: Toronto ; New Jersey : Apple Academic Press, [2015] | Includes bibliographical references and index.
Identifiers: LCCN 2015045512 (print) | LCCN 2015048316 (ebook) | ISBN 9781771883184 (hardcover : alk. paper) | ISBN 9781771883191 ()
Subjects: LCSH: Urban ecology (Sociology) | Human ecology. | Climatic changes--Effect of human beings onl. | Sustainable urban development. | City planning--Environmental aspects.
Classification: LCC HT241 .S867 2015 (print) | LCC HT241 (ebook) | DDC 307.1/216--dc23
LC record available at http://lccn.loc.gov/2015045512

Apple Academic Press also publishes its books in a variety of electronic formats. Some content that appears in print may not be available in electronic format. For information about Apple Academic Press products, visit our website at **www.appleacademicpress.com** and the CRC Press website at **www.crcpress.com**

ABOUT THE EDITOR

KIMBERLY ETINGOFF

Kim Etingoff has Tufts University's terminal master's degree in Urban and Environmental Policy and Planning. Her recent experience includes researching with Initiative for a Competitive Inner City a report on food resiliency within the city of Boston. She worked in partnership with Dudley Street Neighborhood Initiative and Alternatives for Community and Environment to support a community food-planning process based in a Boston neighborhood, which was oriented toward creating a vehicle for community action around urban food issues, providing extensive background research to ground the resident-led planning process. She has worked in the Boston Mayor's Office of New Urban Mechanics, and has also coordinated and developed programs in urban agriculture and nutrition education. In addition, she has many years of experience researching, writing, and editing educational and academic books on environmental and food issues.

CONTENTS

ACKNOWLEDGMENT AND HOW TO CITE

The editor and publisher thank each of the authors who contributed to this book. The chapters in this book were previously published elsewhere. To cite the work contained in this book and to view the individual permissions, please refer to the citation at the beginning of each chapter. Each chapter was read individually and carefully selected by the editor; the result is a book that provides a multiperspective look at research into many elements of what makes cities sustainable. The chapters included examine the following topics:

- Chapter one presents the concept of "biophilic cities," urban areas that bring people into close contact with nature on a daily basis, and seeks to explain how biophilic urbanism leads to sustainability and resilience.
- Chapter two focuses on the ways in which social-ecological systems shape urban ecosystem services, calling for greater local-level stewardship.
- Chapter three discusses the issue of integration of climate change policy within other policy agendas by reviewing and comparing the regulative environment in three Nordic cases: adaptation to climate change at the municipal level in Denmark and Finland, and a case of local level mitigation in Norway.
- Chapter four argues that cities should account not only for carbon emissions actually emitted within their boundaries, but also for emissions that result from urban consumption patterns of specific goods. It then proposes an approach to accounting for these emissions using solid waste life cycle assessments.
- The authors of chapter five assess whether larger cities are necessarily more energy and emissions efficient, and find that larger cities produce more emissions than smaller ones, and have not benefited from economies of scale in terms of energy efficiency.
- Chapter six emphasizes the need to understand not only how urban climate change experiments are made and assembled, but also how they are maintained within specific urban contexts.
- To counter unsustainable soil sealing practices, chapter seven assesses more sustainable management practices meant to preserve ecosystem services and maintain and increase urban green space. It finds promising strategies being implemented in European settings, and calls for better implementation.

- Chapter eight introduces a novel indicator and geographic information system-based method for measuring sustainability levels at a variety of scales, meant to be used by urban planning practitioners.
- The research in chapter nine quantifies ground transportation costs to users, incorporating costs of multi-modal users on traffic patterns. It concludes that increasing low-cost infrastructure for bicycles results in benefits for all transportation users.
- Chapter ten discusses several sustainable technologies available for use in peri-urban areas. Based on a participatory research model, concept scenarios were created and analyzed, which demonstrate that the identified technologies can successfully supply adequate water and recover resources for residents.
- Chapter eleven analyzes urban water supply vulnerability in seventy cities and concludes that a large portion are vulnerable today with greater vulnerability by 2040, even under potential mitigation scenarios.

LIST OF CONTRIBUTORS

Erik Andersson
Stockholm Resilience Centre, Stockholm University

Martina Artmann
Department of Geography and Geology, Paris-Lodron-University Salzburg, Hellbrunnerstraße 34, 5020 Salzburg, Austria

Nina Baron
Department of Environmental Science–Enviromental Social Science, Aarhus University, Frederiksborgvej 399, Roskilde 4000, Denmark

Stephan Barthel
Stockholm Resilience Centre, Stockholm University

Timothy Beatley
Department of Urban and Environmental Planning, University of Virginia, Charlottesville, VA 22904, USA.

Iemke Bisschops
Lettinga Associates Foundation, Bornse weilanden 9, 6700 AM Wageningen, The Netherlands

Sara Borgström
Stockholm Resilience Centre, Stockholm University

Vanesa Castán Broto
Development and Planning Unit, Bartlett Faculty of the Built Environment, University College London, London, UK

Harriet Bulkeley
Geography Department, Durham University, Durham, UK

Johan Colding
The Beijer Institute, Royal Swedish Academy of Sciences

Andres F. Clarens
Department of Civil and Environmental Engineering, University of Virginia, Thornton Hall B228, 351 McCormick Road, PO Box 400742, Charlottesville, VA 22904-4742, USA

Fatih Dur
School of Urban Development, Queensland University of Technology, 2 George Street, Brisbane, Queensland 4001, Australia

Thomas Elmqvist
Stockholm Resilience Centre, Stockholm University

Laura Essl
Centre for Environmental Management and Decision Support, Vienna 1180, Austria

Carl Folke
Stockholm Resilience Centre, Stockholm University; The Beijer Institute, Royal Swedish Academy of Sciences

Michail Fragkias
Department of Economics, Boise State University, Boise, Idaho, United States of America

Håkon Fyhn
Department of Interdisciplinary Studies of Culture, Norwegian University of Science and Technology, Trondheim 7491, Norway

Steven M. Gorelick
Department of Environmental Earth System Science, Stanford University, 473 via Ortega, Stanford, CA 94305, USA

Conrad A. Gosse
Department of Civil and Environmental Engineering, University of Virginia, Thornton Hall B228, 351 McCormick Road, PO Box 400742, Charlottesville, VA 22904-4742, USA

Åsa Gren
The Beijer Institute, Royal Swedish Academy of Sciences

Sirkku Juhola
Department of Built Environment, Aalto University, P.O. Box 11000, Aalto FI-00076, Finland; Department of Environmental Sciences, University of Helsinki, P.O. Box 65, Helsinki FI-00014, Finland

E. Carina H. Keskitalo
Department of Geography and Economic History, Umeå University, Umeå 901 87, Sweden

Meidad Kissinger
Department of Geography and Environmental Development, Ben Gurion University of the Negev, Beer Sheva, Israel.

Johannes Klein
Department of Built Environment, Aalto University, P.O. Box 11000, Aalto FI-00076, Finland; Unit of Environmental Geology, Geological Survey of Finland, P.O. Box 96, Espoo 02151, Finland

José Lobo
School of Sustainability, Arizona State University, Tempe, Arizona, United States of America

Eduardo López
The Mexican Institute of Water Technology (IMTA), Cuernavaca 62550, México

José Luis Martínez-Ruiz
The Mexican Institute of Water Technology (IMTA), Cuernavaca 62550, México

Jennie Moore
Department of Geography and Environmental Development, Ben Gurion University of the Negev, Beer Sheva, Israel.

Daniel Murillo
The Mexican Institute of Water Technology (IMTA), Cuernavaca 62550, México

Tiemen A. Nanninga
Lettinga Associates Foundation, Bornse weilanden 9, 6700 AM Wageningen, The Netherlands

Peter Newman
Curtin University Sustainability Policy (CUSP) Institute, Curtin University, Fremantle, WA 6160, Australia.

Julie C. Padowski
Department of Environmental Earth System Science, Stanford University, 473 via Ortega, Stanford, CA 94305, USA

William E. Rees
Department of Geography and Environmental Development, Ben Gurion University of the Negev, Beer Sheva, Israel.

Markus Starkl
Competence Centre for Decision-Aid in Environmental Management, University of Natural Resources and Life Sciences/DIB, Gregor Mendel Strasse 33, Vienna 1180, Austria

Deborah Strumsky
Department of Geography & Earth Sciences, University of North Carolina-Charlotte, Charlotte, North Carolina, United States of America

Cornelia Sussmann
Department of Geography and Environmental Development, Ben Gurion University of the Negev, Beer Sheva, Israel.

Tan Yigitcanlar
School of Urban Development, Queensland University of Technology, 2 George Street, Brisbane, Queensland 4001, Australia

INTRODUCTION

Two trends come together in the world's cities to make urban sustainability a critical issue today. First, greater and greater numbers of people are living in urban areas, and are projected to do so for the foreseeable future. Additionally, cities contribute to climate change in a significant way, and must make systemic changes to mitigate and adapt to climate change effects. Urban planners face serious challenges in enhancing sustainability, but also have an important set of tools available for creating innovative solutions. This book adds to the conversation about the place of urban planning in the creation and maintenance of sustainable cities.

Urban sustainability has a variety of dimensions. The first section of this book connects sustainability and the role of nature and ecosystem services in a healthy urban setting. The second part specifically looks at climate change and carbon emissions in cities, and particularly advances an understanding of the relationship between urban areas, energy, and carbon. The third section includes chapters on measuring and advancing the sustainability of land-use and infrastructure such as transportation. Finally, the fourth section takes a closer look at one of the most pressing sustainability issues, the availability of water for urban dwellers around the world.

—*Kimberly Etingoff*

There is a growing recognition of the need for daily contact with nature in order to live happy, productive, meaningful lives. Recent attention to biophilic design among architects and designers acknowledges this power of nature.

However, in an increasingly urban planet, more attention needs to be aimed at the urban scales, at planning for and moving towards what the authors of chapter 1 call "biophilic cities." Biophilic cities are cities that provide close and daily contact with nature, nearby nature, but also seek to foster an awareness of and caring for this nature. Biophilic cities, the authors argue, are also sustainable and resilient cities. Achieving the conditions of a biophilic city will go far, they say, in helping to foster social and landscape resilience, in the face of climate change, natural disasters, economic uncertainty, and various other shocks that cities will face in the future. Their article identifies key pathways by which biophilic urbanism enhances resilience. While some are well-established relationships, others are more tentative and suggest future research and testing.

Within-city green infrastructure can offer opportunities and new contexts for people to become stewards of ecosystem services. In chapter 2, the authors analyze cities as social-ecological systems, synthesize the literature, and provide examples from more than 15 years of research in the Stockholm urban region, Sweden. Their social-ecological approach spans from investigating ecosystem properties to the social frameworks and personal values that drive and shape human interactions with nature. Their findings demonstrate that social-ecological systems generate urban ecosystem services and that local stewards are critically important. However, land-use planning and management seldom account for their role in the generation of urban ecosystem services. While the small-scale patchwork of land uses in cities stimulates intense interactions across borders, much focus is still on individual patches. The authors' results highlight the importance and complexity of stewardship of urban biodiversity and ecosystem services and of the planning and governance of urban green infrastructure.

Recently, considerable focus, e.g., in the fifth IPCC (Intergovernmental Panel on Climate Change) Assessment Report (2014) has been trained on why adaptation and mitigation have not been developed more than at present, with relatively few local government actions taken compared with, for example, more discursive policy agreement on the importance of the issue of climate change. Going beyond a focus on general limits and barriers, the authors of chapter 3 suggest that one important issue is that climate change has not yet been sufficiently integrated into the state regulative structure of legislation and policy-making. A comparison between three cases suggests that local developments that are not supported in particular by binding regulation are unlikely to achieve the same general level of implementation as issues for which such regulative demands (and thereby also requirements for prioritization) exist. This constitutes an important consideration for the development of adaptation and mitigation as policy areas, including on the local level.

Although many cities are engaged in efforts to calculate and reduce their greenhouse gas (GHG) emissions, acknowledge the authors of chapter 4, most are accounting for "scope one" emissions (i.e., GHGs produced within urban boundaries, such as following the protocol of the International Council for Local Environmental Initiatives). The authors state that cities should also account for the emissions associated with goods, services, and materials consumed within their boundaries—"scope three" emissions. The emissions related to urban consumption patterns and choices greatly influence overall emissions that can be associated with an urban area. However, data constraints and GHG accounting complexity present challenges. In this paper, the authors propose one approach that cities can take to measure the GHG emissions of their material consumption: the solid waste life cycle assessment (LCA) approach. They use this approach to identify a set of materials commonly consumed within cities, and reviewed published life cycle assessment data to determine the GHG emissions associated with production of each. Their review reveals that among fourteen commonly consumed materials, textiles, and aluminum are associated with the highest GHG emissions per ton of production. Paper and plastics have relatively lower production emissions, but a potentially higher impact on overall emissions owing to their large proportions, by weight, in the consumption stream.

Urban areas consume more than 66 percent of the world's energy and generate more than 70 percent of global greenhouse gas emissions. With the world's population expected to reach 10 billion by 2100, nearly 90 percent of whom will live in urban areas, a critical question for planetary sustainability is how the size of cities affects energy use and carbon dioxide (CO_2) emissions. Are larger cities more energy and emissions efficient than smaller ones? Do larger cities exhibit gains from economies of scale with regard to emissions? In chapter 5, the authors examine the relationship between city size and CO_2 emissions for U.S. metropolitan areas using a production accounting allocation of emissions. They find that for the time period of 1999–2008, CO_2 emissions scale proportionally with urban population size. Contrary to theoretical expectations, larger cities are not more emissions efficient than smaller ones.

Climate change governance is increasingly being conducted through urban climate change experiments, purposive interventions that seek to reconfigure urban sociotechnical systems to achieve low-carbon and resilient cities. In examining how experiments take effect, the authors of chapter 6 suggest that we need to understand not only how they are made and assembled, but also how they are maintained within specific urban contexts. Drawing on literatures from urban political ecology and the specific debate on urban repair and maintenance, this chapter examines maintenance in two case studies of climate change experiments in housing in Bangalore (India) and Monterrey (Mexico). The authors find that maintenance is a crucial process through which not only urban obduracy is preserved, but also the novel and innovative character of the experiment is asserted and reproduced. The process of "maintaining" experiments is a precarious one, which requires a continuous external input in terms of remaking the experiment materially and discursively. This process causes further reconfigurations beyond the experiment, changing the patterns of responsibility attribution and acceptability that configure the urban fabric.

Chapter 7 turns our attention to other urban issues. Soil sealing has negative impacts on ecosystem services, since urban green and soil get lost. Although there is political commitment to stop further sealing, no reversal of this trend can be observed in Europe. This paper raises these questions: (1) Which strategies can be regarded as being efficient toward ecologically sustainable management of urban soil sealing? (2) Who has competences and should take responsibility to steer soil sealing? The analyses are conducted in Germany, and the assessment of strategies is carried out using indicators as part of a content analysis. Legal-planning, informal-planning, economic-fiscal, co-operative, and informational strategies are analyzed. Results show that there is a sufficient basis of strategies to secure urban ecosystem services by protecting urban green and reducing urban gray where microclimate regulation is a main target. However, soil-sealing management lacks a spatial strategically overview as well as the consideration of services provided by fertile soils.

Measuring the comparative sustainability levels of cities, regions, institutions and projects is an essential procedure in creating sustainable urban futures. Chapter 8 introduces a new urban sustainability assessment model: "The Sustainable Infrastructure, Land-use, Environment and Transport model" (SILENT). The SILENT model is an advanced geographic information system and indicator-based comparative urban sustainability indexing model. The model aims to assist planners and policy makers in their daily tasks in sustainable urban planning and development by providing an integrated sustainability assessment framework. Chapter 7 gives an overview of the conceptual framework and components of the model and discusses the theoretical constructs, methodological procedures, and future development of this promising urban sustainability assessment model.

Efforts to reduce the environmental impacts of transportation infrastructure have generally overlooked many of the efficiencies that can be obtained by considering the relevant engineering and economic aspects as a system. In chapter 9, the authors present a framework for quantifying the burdens of ground transportation in urban settings that incorporates travel time, vehicle fuel, and pavement maintenance costs. A Pareto set of bi-directional lane configurations for two-lane roadways yields non-dominated combinations of lane width, bicycle lanes, and curb parking. Probabilistic analysis and microsimulation both show dramatic mobility reductions on road segments of insufficient width for heavy vehicles to pass bicycles without encroaching on oncoming traffic. This delay is positively correlated with uphill grades and increasing traffic volumes and inversely proportional to total pavement width. The response is nonlinear with grade and yields mixed uphill/downhill optimal lane configurations. Increasing bicycle mode share is negatively correlated with total costs and emissions for lane configurations allowing motor vehicles to safely pass bicycles, while the opposite is true for configurations that fail to facilitate passing. Spatial impacts on mobility also dictate that curb parking exhibits significant spatial opportunity costs related to the total cost Pareto curve. The proposed framework provides a means to evaluate relatively inexpensive lane reconfiguration options in response to changing modal share and priorities. These results provide quantitative evidence that efforts to reallocate limited pavement space to bicycles, like those being adopted in several US cities, could appreciably reduce costs for all users

Often centralized water supply, sanitation, and solid waste services struggle to keep up with the rapid expansion of urban areas. The peri-urban areas are at the forefront of this expansion, and it is here where decentralized technologies are increasingly being implemented. The introduction of decentralized technologies allows for the development of new opportunities that enable the recovery and reuse of resources in the form of water, nutrients and energy. This resource-oriented management of water, nutrients, and energy requires a sustainable system aimed at low resource use and high recovery and reuse rates. Instead of investigating each sector separately, as has been traditionally done, chapter 10 proposes and discusses a concept that seeks to combine the in- and outflows of the different sectors, reusing water and other liberated resources where possible. This paper shows and demonstrates examples of different types of sustainable technologies that can be implemented in the peri-urban areas of Mexico City (rainwater harvesting, EcoSan and *biofiltros* [small constructed wetlands], and [vermi-]composting). An innovative participatory planning method, combining scenario development with a participatory planning workshop with key stakeholders, was applied and resulted in three concept scenarios. Specific technologies were then selected for each concept scenario that the technical feasibility and applicability was assessed. Following this, the resulting resource flows (nutrients, water, and energy) were determined and analyzed. The results show that decentralized technologies not only have the potential to deliver adequate water supply, sanitation and solid waste services in peri-urban areas and lessen environmental pollution, but also can recover significant amounts of resources thereby saving costs and providing valuable inputs in, for instance, the agricultural sector. Social acceptance of the technologies and institutional cooperation, however, is key for successful implementation.

Chapter 11 presents a global analysis of urban water supply vulnerability in 71 surface-water supplied cities, with populations exceeding 750,000 and lacking source water diversity. Vulnerability represents the failure of an urban supply-basin to simultaneously meet demands from human, environmental, and agricultural users. The authors assess a baseline (2010) condition and a future scenario (2040) that considers increased demand from urban population growth and projected agricultural demand. They do not, however, account for climate change, which can potentially exacerbate or reduce urban supply vulnerability. In 2010, 35 percent of large cities are vulnerable as they compete with agricultural users. By 2040, without additional measures, 45 percent of cities are vulnerable due to increased agricultural and urban demands. Of the vulnerable cities in 2040, the majority are river-supplied with mean flows so low (1200 liters per person per day, l/p/d) that the cities experience chronic water scarcity (1370 l/p/d). Reservoirs supply the majority of cities facing individual future threats, revealing that constructed storage potentially provides tenuous water security. In 2040, of the 32 vulnerable cities, 14 would reduce their vulnerability via reallocating water by reducing environmental flows, and 16 would similarly benefit by transferring water from irrigated agriculture. Approximately half remain vulnerable under either potential remedy.

PART I

CITIES AND NATURE

CHAPTER 1

Biophilic Cities Are Sustainable, Resilient Cities

TIMOTHY BEATLEY AND PETER NEWMAN

1.1 INTRODUCTION: THE EMERGENCE OF A BIOPHILIC PERSPECTIVE ON CITIES

There is increasing interest on the part of architects, planners and urban designers in biophilic design and much new writing and literature appearing in the last several years. Biophilic design holds that good design, at the building, site, city and regional scale, must include nature and natural elements. It is based especially on the concept of *biophilia*, popularized by Harvard myrmecologist and sociobiologist E.O. Wilson. Wilson argues that humans have co-evolved with nature and that we carry with us our ancient brains and our need to connect with and affiliate with nature, to be happy and healthy. Wilson defines biophilia as "the innately emotional affiliation of human beings to other living organisms. Innate means hereditary and hence part of ultimate human nature". To Wilson, biophilia is

Biophilic Cities Are Sustainable, Resilient Cities. © *2013 by the authors; licensee MDPI, Basel, Switzerland.* Sustainability *2013, 5(8), 3328-3345; doi:10.3390/su5083328. Creative Commons Attribution license (http://creativecommons.org/licenses/by/3.0/).*

really a "complex of learning rules" developed over thousands of years of evolution and human-environment interaction [1,2,3].

And there is now a growing body of evidence of the positive physical and mental health benefits associated with greenery and green elements in living and work environments. Research at the building scale shows strong positive relationships between the presence of natural daylight, fresh air and greenery, with increases in worker happiness and productivity [4]. Similar studies show the positive power of schools that incorporate natural daylight and other green elements, to raise test scores of the students [5]. Considerable evidence and research, dating to the 1980's, has suggested the healing and recuperative power of nature in hospitals and health care facilities (including Ulrich's [6] classic study showing recovery from gall bladder surgery is significantly enhanced by views of nature). Evidence of the power of the green qualities and features is also emerging at neighborhood community scales. Green neighborhoods and more natural living environments have been associated with reductions in stress and increased levels physical and mental health [7,8,9,10,11]. An important study in *The Lancet* concludes that populations with greater exposure to green space experience lower mortality and that green space exposure can help reduce health inequalities [12]. The presence of nature, moreover, is associated with improvements in positive mood, cognitive performance and even creativity [13]. A recent pilot study using portable electroencephalography (EEG) caps further demonstrates the value of nature in reducing mental fatigue [14]. Nature has immense power to restore, heal and fascinate.

This emerging evidence and research has helped in turn to increase interest among urban designers and architects in designing buildings and facilities that enhance nature. While much energy and attention of late has been focused on biophilic design, this has largely assumed a focus on the building or site. Beatley [15] and others argue that while integrating green and natural elements into building design is critical, there is much value in fact in getting people out of buildings and to thinking more holistically about the natural qualities and conditions of the larger urban environments in which these buildings sit.

Cities and urban environments contain a variety of ecological and green assets, from parks to trees to rivers and riparian habitats, and increasingly, efforts are being made to further enhance the green elements and features of these living and work environments. Daylighting urban streams (taking them out of underground pipes and returning them to the surface), installing trails, planting new trees and forests, community gardens, installing green walls and vertical gardens, are among the many ways in which cities and urban environments can become greener. Biophilic urbanism can and must happen at different scales, and Table 1 presents some examples of biophilic design interventions that are possible.

Increasingly cities have developed and are implementing a host of biophilic programs, policies and initiatives. Cities like Chicago and Portland have developed extensive incentives and subsidies for installation of green features, such as green rooftops. Furthermore, increasingly, green features, such as green rooftops, are mandated, as in Toronto, for roofs over a certain size. Some cities, such as Seattle, have established so-called Green Factor standards, mandating minimum green and landscaping elements for certain types of new development, and other cities, such as Chicago, Baltimore and Montreal, are encouraging the greening of alleyways and other otherwise grey spaces in the city. Many cities have established extensive treeing programs and set ambitious tree-planting goals, with the cities of New York, Los Angeles and Houston each setting the goal of a million new trees. Furthermore, many American cities, such as Chicago and San Francisco, have modified their planning and zoning codes to permit urban agriculture.

The physical environment of cities, then, represents an essential requisite for creating biophilic cities. However, the true extent to which a city and its residents can be said to be biophilic, will depend on many other things, including whether and the extent to which citizens avail themselves of this nearby nature and the amount of time residents actually spend out-of-doors. How much they know about and care about this nearby nature is also an important indicator. A biophilic city, moreover, is also a city in which residents are actively involved in experiencing nature—e.g., hiking, bird watching, sky-gazing, gardening, among many other activities. Furthermore, citizens in biophilic cities have abundant opportunities to be engaged in restoring and caring for the nature around them. Table 2 presents a more comprehensive listing of the key qualities, not just those of physical design, by which a biophilic city might be described or defined (for more detail, see Beatley [15]). It is important to recognize that biophilic

Table 1. Biophilic city design elements across scales. Modified from Girling and Kellett [16]; first appeared in Beatley [15].

Scales	Biophilic design elements
Building	Green rooftops
	Sky gardens and green atria
	Rooftop garden
	Green walls
	Daylit interior spaces
Block	Green courtyards
	Clustered housing around green areas
	Native species yards and spaces
Street	Green streets
	Urban trees
	Low impact development (LID); Vegetated swales and skinny streets
	Edible landscaping
	High degree of permeability
Neighborhood	Stream daylighting, stream restoration
	Urban forests
	Ecology parks
	Community gardens
	Neighborhood parks/pocket parks
	Greening greyfields and brownfields
Community	Urban creeks and riparian areas
	Urban ecological networks
	Green schools
	City tree canopy
	Community forest/community orchards
	Greening utility corridors
Region	River systems/floodplains
	Riparian systems
	Regional greenspace systems
	Greening major transport corridors

cities are not simply green cities. The presence of abundant nature is a necessary, but not sufficient condition, and the "philic" is as important as the bio. In biophilic cities, residents are directly and actively engaged in learning about, enjoying and caring for the nature around them and have developed important emotional connections with this nature.

1.2 BIOPHILIC CITIES AND URBAN SUSTAINABILITY AND RESILIENCE

Sustainability has emerged over the last two decades especially as an important goal and frame of reference for cities, and both authors have written extensively about what urban sustainability entails [17,18]. Sustainability is ideally understood as a holistic frame of reference for guiding city development and for helping cities to do many things at once: to reduce their ecological footprints and resource needs, to deepen connections to landscape and place and to enhance livability and quality of life while expanding economic opportunities for the least-advantaged, among others.

Resilience has emerged as another important parallel concept and goal and an urban aspiration increasingly stated alongside (and sometimes replacing) sustainability. Given the impacts (current and potential) of global climate change, an increasingly volatile climate and the already serious range of disasters and hazards faced by cities around the world, global resource conflicts and constraints, long term decline in global oil supply and a global economic system that seems increasingly susceptible to vicissitudes and flux, resilience resonates well as a concept and goal, and we consider it a potent version or flavor of urban sustainability. What began with adaptation to disasters and hazards (many of us began speaking in terms of disaster-resistant and, later, disaster-resilient communities) has now broadened to discussions of resilient cities, that take into account the fuller range of potential shocks and stresses cities will likely face in the future, from water scarcity, rising food prices and higher summer temperatures [19]. Resilience has many meanings, of course, but at its core is the essential ability to successfully adapt to and respond to these shocks; the word derives from the Latin *resiliere*, meaning to jump back or rebound. Godschalk, while writing primarily from a natural hazards perspective, describes a resilient city as one that "would be capable of withstanding severe shock without either immediate chaos or permanent harm … While

Table 2. Some important dimensions of biophilic cities (and some possible indicators thereof). Summarized from Beatley [15].

Biophilic Conditions and Infrastructure
—Percentage of population within a few hundred feet or meters of a park or greenspace;
—Percentage of city land area covered by trees or other vegetation;
—Number of green design features (e.g., green rooftops, green walls, rain gardens);
—Extent of natural images, shapes, forms employed in architecture and seen in the city;
—Extent of flora and fauna (e.g., species) found within the city;
Biophilic Behaviors, Patterns, Practices, Lifestyles
—Average portion of the day spent outside;
—Visitation rates for city parks;
—Percent of trips made by walking;
—Extent of membership and participation in local nature clubs and organizations;
Biophilic Attitudes and Knowledge
—Percent of residents who express care and concern for nature;
—Percent of residents who can identify common species of flora and fauna;
Biophilic Institutions and Governance
—Priority given to nature conservation by local government; percent of municipal budget dedicated to biophilic programs;
—Existence of design and planning regulations that promote biophilic conditions (e.g., mandatory green rooftop requirement, bird-friendly building design guidelines);
—Presence and importance of institutions, from aquaria to natural history museums, that promote education and awareness of nature;
—Number/extent of educational programs in local schools aimed at teaching about nature;
—Number of nature organizations and clubs of various sorts in the city, from advocacy to social groups.

they might bend from hazards forces, they would not break. Composed of networked social communities and lifeline systems, resilient cities would become stronger by adapting to and learning from disasters" [20,21,22]. Resilience does not imply a return to dysfunctional or unsustainable community conditions, but adaptation to dynamic social and ecological conditions in ways that protect and enhance quality of life, long term ecological productivity and public and personal health.

Resilience is not only a matter for cities and regions, of course, but has important implications for individuals and families. It is here often where the stresses, pressures and shocks of modern life will come to bear. Can individuals and families adapt and cope and in what ways? Will individuals and families be able to effectively and adequately adapt to the hotter conditions of urban neighborhoods, for instance, a likely result of climate change? Will they be able to respond effectively to future natural disaster events, as well as the impacts of future economic downturns (e.g., unemployment, rising food prices)? We are concerned here as well with the steps that can be taken to foster resilience at these levels: the ability of individuals and families to spring-back, to adapt and to take advantage of the opportunities to enhance their health and wellbeing.

It is the key premise of this paper that there are important relationships between *biophilia* or *biophilic cities* and urban sustainability and resilience and, more specifically, that the former helps to advance the latter. That is movement in the direction of making cities greener, more natural, more *biophilic*, will also help to make them more resilient. There are many pathways from biophilic design and urban biophilia to urban resilience, many ways in which the conditions of green and biophilic cities will also serve to make a city more resilient in the long run, ecologically, economically and socially. Some of these biophilic pathways are direct: as when investments in green infrastructure (say restoring wetlands or planting drought tolerant vegetation in cities) serves to yield resilience benefits and outcomes (e.g., reduced summer temperatures, reduced flooding from coastal storms). Other pathways are more indirect: as when green elements serve to stimulate or enhance beneficial and health-inducing behaviors, such as walking, that in turn serves to enhance the resilience of individuals and families to cope with future stresses. Figure 1 presents a very tentative conceptual diagram depicting the general flow of these pathways and suggesting some especially important pathways through which biophilic cities or biophilic urbanism, might positively influence urban resilience.

1.3 SOME KEY PATHWAYS FROM BIOPHILIC URBANISM TO URBAN RESILIENCE

1.3.1 RESILIENCE THROUGH URBAN NATURAL SYSTEMS

There are a number of primary pathways, we believe, by which biophilic cities or biophilic urbanism will enhance or increase urban resilience. Some of these pathways and relationships have been well-researched and well-established. Others are more tentative and in need of additional work.

One of the clearest pathways is biophysical, the resilience benefits provided through the protection and enhancement of the natural systems and features in and around a city. We know that the green infrastructure of a city and region—rivers and riparian areas, floodplains and wetlands and large swaths of forested land—all provide essential services, that help cities and urban regions respond to and spring back from climatic and natural events. Cities with large natural wetland systems will be better able to absorb flood waters from hurricanes and storms, for instance. In New Orleans, for instance, vulnerability to storms and flooding has been significantly increased as a result of a long history of wetlands alterations. Costanza, Mitsch and Day [23] have argued compellingly for giving renewed importance to preserving and protecting an intact system of wetlands around New Orleans as a key strategy for long term resilience. These flood protection services carry a considerable economic value that must be recognized: "Had the original wetlands been intact and levees in better shape, a substantial portion of the US$100 billion plus damages from this hurricane [Katrina] probably could have been avoided" [23]. In a more recent paper, Costanza and colleagues estimate that coastal wetlands provide some $23 billion a year in hurricane protection benefits [24]. It has been estimated that every mile of protected forested cypress swamp in Louisiana, reduces flood surge heights by a foot or more. Protecting and

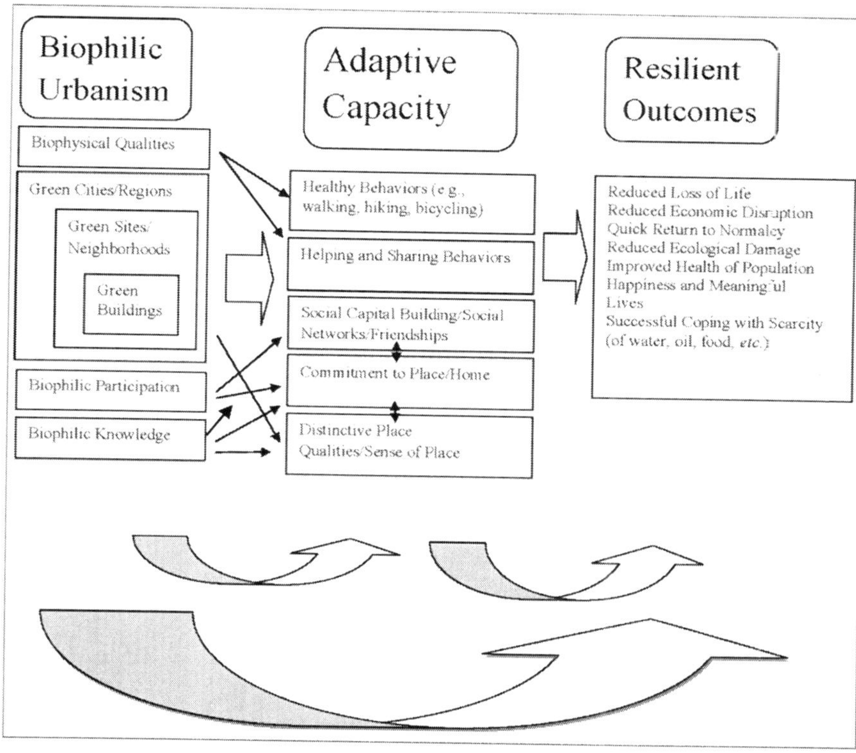

Figure 1. Biophilic pathways to urban resilience.

restoring these larger ecological systems is a wise move in becoming more disaster-resilient.

Recent planning and policy initiatives around water and water supply for cities further suggest the direct resilience benefits of green infrastructure for enhancing water supply and water system resilience. The story of New York City's investment in the conservation and protection of its upper watersheds in the Catskills Mountains, is suggestive of the power of this framework. Here, the city has invested in acquisition and long term management of land (more than 70,000 acres protected through fee-simple or less-than-fee simple acquisition), a far more cost-effective method than the multi-billion dollar costs of water filtration plants and a model strongly supporting green or ecological infrastructure [25].

More recently, cities such as Denver, have entered into new long term restoration and conservation arrangements to make the city less susceptible to the water-supply impacts associated with wildfire and forest degradation in its source watersheds. The initiative called Forests to Faucets, more specifically, is an agreement between Denver Water and the U.S. Forest Service, an acknowledgement that the City of Denver's water supply has been seriously compromised in recent years as a result of wildfire-related sediment (the 2002 Hayman Fire, for instance, resulted in the deposition of more than 1 million cubic yards of sediment into one of the city's main water supply reservoirs). The agreement calls for a shared funding for a comprehensive program of interventions to enhance forest health and to address the problems of infestations from insects and threats of wildfire. These steps together will protect water supply delivery system for the city of Denver and will, as well, serve to make the forest upstream of the city more resilient in the face of these ecosystem changes. As the Denver water press release notes: "Restoration also will help the forests become more resistant to future insect and disease, reduce wildfire risks and maintain habitat for fish and wildlife. More resilient forests will also be more adaptive to the impacts of a changing climate" [26].

Cities with extensive tree canopy coverage provide many ecological benefits that will make cities more resilient—including moderation of air pollutants, cooling through evapo-transpiration and shading and reduction in urban flooding and runoff. Protection and restoration of urban streams and rivers will reduce vulnerability again to floods, provide important cooling benefits and help to moderate the weather and temperature changes predicted as a result of climate change (e.g., east coast American cities are predicted to have substantially higher summer highs by 2050). The cumulative impacts of green features, such as green rooftops and vertical gardens, can be significant indeed. In Toronto, the first North American city to now mandate installation of green rooftops for roofs over a certain size, estimates suggest that conversion of flat roofs to green rooftops would reduce urban temperatures by 1.5 degree Celsius or more [27].

Cities with more extensive networks of parks and greenspaces (though their design and configuration will matter) are also likely to fare better in the face of long term climate change. The City of Brisbane, Australia, a remarkably green city, aspires to create a connected network of natural areas, with topographic variation and diversity that will allow native biodiversity to adapt and shift as climate changes [28].

Trees and natural vegetation in cities and urban neighborhoods can help protect property and reduce damage from wind, rain and flooding. Some communities, such as Charleston County, SC, have developed programs for encouraging "hazard-resistant landscaping", planting native trees, such as live oaks, that tend to fare well in the face of high winds [29]. Working to make a neighborhood or community's landscape more resilient means finding ways to insert trees, landscaping, green rooftops and use of water sensitive urban design features, such as bio-swales and rain gardens.

Biophilic urbanism can help to protect or strengthen favorable climatic and micro-climate conditions in cities. A number of European cities, notably German cities, have a long history of spatial planning with climate in mind. Notable examples include Freiburg and Stuttgart, both well-known for their efforts at protecting natural air movement into and out of the city, as a way of addressing urban heat and enhancing air quality and generally the healthfulness and pleasantness of city life. Freiburg, for instance, imposes height limitations in certain zones to prevent interruption of these air flows and also designs new buildings and neighborhoods to permit better flow-through. Stuttgart, as a further example, has designated a series of "ventilation lanes" where future development is restricted [30].

Many of the biophilic city elements contained in Table 1 and Table 2, will also help cities become more resilient in the face of a host of emerging resource scarcities likely in the decades ahead, including long term decline in global oil supply (peak oil), diminished supplies of potable water and food, among others. Greening cities can, as already mentioned, significantly reduce energy consumption and reduce consumption of energy for heating and cooling. In promoting modes of urban mobility other than automobiles (walking, bicycling) there is the possibility of greater resilience in the diminished oil supplies. Equally true, biophilic urbanism can achieve significant water conservation and, through the protection of peri-urban farms and agriculture and by promoting urban agriculture, might help to ensure the food security of a city.

As Figure 1 suggests, these natural features and urban biophilic qualities directly help cities to become more resilient, but also serve more indirectly to encourage healthier lifestyles and long term coping behaviors that will stand communities in good stead in the face of many future shocks and stresses. Trees and green elements at the neighborhood level, for instance, have been shown to encourage walking and outdoor-oriented activity and, in this way, create conditions for healthier individuals and families and laying a foundation for urban resilience [31].

1.3.2 BIOPHILIC CITIES CAN HELP TO STRENGTHEN COMMITMENTS TO PLACE

Urban environments that are greener, more nature-full, will attract greater interest by residents and help to strengthen emotional bonds to place and community, in turn increasing urban resilience [32].

These green elements are an important aspect of place. They serve as important amenities that can strengthen distinctive place qualities, nurturing a special sense of place and, in turn, commitments to place [32,33]. When identifying those cities where there is a strong sense of the unique, that look and feel differently, elements of the natural environment are critical. Some of these, such as topography and climate, are essentially givens, that a city can do little about, but there are many ways in which through creative planning and development, connections to these elements can be strengthened, with the concomitant result that an urban citizenry may exhibit a stronger emotional connection with and care for the landscape and cityscape they call home. Furthermore, this, in turn, while a tentative pathway in need of additional research and evidence, may result in a host of other resilience-enhancing actions, attitudes and behaviors. Places viewed as special and distinct will likely foster emotional bonds and commitments that may in turn translate into a variety of conditions helpful to resilience (e.g., a citizenry willing to stay put and to roll up their sleeves to rebuild following a disaster or a citizenry politically engaged and participative, for instance).

A dramatic example can be seen in Austin, Texas, a city that has emphasized the cultivating of qualities that make it different or unusual (and even a buy-local campaign dubbed "Keep Austin Weird"). One of the more impressive ways that Austin has solidified its special sense of place is through the natural environment and most dramatically the way in which this city has now embraced its part-time resident population of Mexican free-tailed bats. A million and a half of these bats take up residence each spring and summer under the city's Congress Avenue bridge, and lining along the top of the bridge or sitting on blankets below the bridge to watch the nightly emergence of the bats has become a tradition in the summer and a significant tourist draw, attracting some 100,000 tourists to Austin each year. The bats deliver many ecological benefits, of course, and they bring considerable direct economic benefits to the city, but they also are part of what makes Austin different, or special, and are an important part of the unique character and feel of the city.

These biophilic qualities are an important part of what makes this city, and others, a place that Richard Florida argues is a city attractive to the so-called *creative class* [33]. There are thus considerable economic benefits from these biophilic programs and elements, as their distinctive place contributions serve to enhance quality of life and in turn the attractiveness of these cities. Whether cities like Austin will be able to retain businesses and residents in the face of economic stresses and downturns, and whether cultivating distinctive place qualities will deliver positive competitive advantages that might help such biophilic cities out-compete other similar cities remain open questions. However, there is considerable anecdotal evidence that cities that are greener, have more extensive green amenities, from trail systems to extensive parks to access to nature, will fare better in difficult global economic times. Furthermore, many cities have mayors and leadership teams that support improvements in biophilic conditions as an essential approach to economic development (e.g., now former mayor Richard Daley of Chicago).

Attachment and commitment to place represent significant precursors, it would seem, for taking (or being prepared to take) significant steps on behalf of a community in response to a crisis or future urban shock [34]. However, the extent to which, and precise ways in which, a unique sense of place, and personal attachments and commitments to place, will result in greater urban resilience remain open questions. A stronger commitment to place can be hypothesized to lead to a host of actions that might serve to enhance resilience and might correlate with resilience behaviors.

1.3.3 BIOPHILIC CITIES CAN ENHANCE FAMILY AND INDIVIDUAL RESILIENCE AND ADAPTIVE CAPACITY

Biophilic cities deliver a host of positive benefits at an individual and family level that will help to enhance ability to successfully cope with and adapt to future stressors and shocks. Nature in cities delivers considerable and often underappreciated health benefits, furthering bolstering city's capacity and ability to adapt to stresses and shocks in the future. Residents of greener neighborhoods have been found to walk more and spend more time outside, in turn with considerable positive health effects [35].

And in times of individual and family stress, parks, green spaces, oases of nature, become even more important. As the 2008 global economic downturn unfolded, there is evidence that urban park usage went up; nature delivers importance solace and emotional sustenance, at precisely the times when individuals and families need them [36]. A city with few such green spaces, will be less prepared, less able to provide such emotional comfort.

The evidence is compelling, as mentioned above, that the natural elements of cities—even views of trees and nature—deliver important positive impacts on mood and reduction of fatigue and, in turn, enhancing individual resilience and health. A recent study by the UK mental health organization MIND, sought to gauge the positive mood effects of a walk in nature, and found in comparison to walking in an indoor structure, the impacts were considerable indeed. At the end of these walks, participants showed marked decreases in depression, anger, tension, confusion and fatigue and increases in vigor [9].

And there are many more direct ways, of course, in which green urbanism and biophilic urban design can create more healthful environments. Planting trees and installing green rooftops and walls, permeable paving, are all strategies that provide shade and cooling through evapo-transpiration. These help to moderate urban temperatures. Trees help to ameliorate air pollution, and a robust and extensive network of storm water collection features (from permeable paving to rain gardens to urban forests) helps to reduce the problem of combined sewer overflow (CSO), a serious water quality problem that many American cities, from Seattle to Washington DC, are afflicted with.

Biophilic urbanism can help to provide the basis for healthier lives and lifestyles in many ways. An extensive network of walking and hiking trails, the close physical proximity of large blocks of greenspaces to urban neighborhoods and an urban land use pattern of nearby parks, all contribute to at the least the possibility of a more physically active, healthier lifestyle. Gardening and food production opportunities in the city provide similar benefits—the opportunity to engage in an emotionally rewarding activity, but also one with substantial calorie-burning potential and the possibility of improving diets, as well [37].

There is special concern about the unhealthy lives of children, as seen in rising child obesity rates, unhealthy diets and lack of physical activity, also little connection with outside nature. Richard Louv, in his groundbreaking book *Last Child in the Woods*, dubs the latter phenomenon "nature deficit disorder", pointing to many culprits of modern life, including elevated concerns about safety of outdoor environments, the overdependence on electronic media and the criminalizing of outdoor play, among others ([38]; there is a growing body of research on the importance of nature in childhood health and development, see [39,40]). The time spent by children on electronic media has actually increased in recent years [41], boding ill for the kinds of contact with the natural world that will foster a lifelong love of and comfort and wonder provided by nature, as well as the physical exercise and activity that outside play generates.

Biophilic cities, as Table 2 indicates, go beyond simply provision of parks and nearby nature, but seek to actively entice and induce urbanites to enjoy and participate in this nature. Cities like New York, have taken a variety of creative steps. For instance, during the summer months of July and August families can camp in city parks throughout the five boroughs. The city's Parks Department website emphasizes the fun families will have: "We start with a cookout, then a variety of evening activities, such as stargazing and night hikes. You'll sleep under the stars and awake to the birds" [42].

Delivering the raw ingredients of wonder and fascination in cities is a key goal of biophilic urbanism. Partly this involves re-imagining cities of abundant life and actively exploring ways that human urbanites can co-exist with other animals, from the Mexican free-tailed bats mentioned above to coyotes, birds and invertebrates, all biodiversity that cities often have in great abundance, though could do more to protect and nurture. Viewing birds (and listening to their calls) and watching other wildlife, at once provides mental and emotional connections, stress reduction and other biophysical benefits [43,44,45]. We find much to recommend in Jennifer Wolch's concept of "zoöpolis", arguing for the need to actively cultivate and accommodate animals in urban environments and in the process "re-enchanting" our cities [46]. Biophilic cities are cities that accommodate and celebrate other life forms and in the process help to advance resilience. Often the animals and wildlife of a city are an important contributor to a unique or distinctive sense of place [15].

Nature imparts many other affective benefits to urbanites that we are just beginning to better understand. Recent experimental evidence suggests that humans are more generous in the presence of nature [47]. Generosity and a willingness to engage in a variety of helping behaviors, are important background conditions in the event of a need to respond to and recovery from a serious natural disaster, such as an earthquake or hurricane, for instance. The sharing of critical information in the situation of an approaching storm requires a certain commitment to and concern about the welfare of others, personal qualities that may be significantly more evident in green urban environments.

Regular participation in activities that engage nature in cities can have many positive emotional and physiological benefits. Evidence is irrefutable that watching birds and wildlife can reduce stress, lower heart rates, help to calm and, in medical settings, heal us. There is much emphasis of late in the architectural community about the importance of hospital and medical facilities that integrate natural daylight, healing gardens and other features of nature, and this is a positive trend, to be sure. However, in many ways it is the larger potential of designing and planning healing cities and urban neighborhoods that is most exciting, spaces and environments that might exert considerable preventative benefits (and preventing the need to visit that hospital or cancer center later in life, it could be hoped).

Food production in cities offers opportunities to at once reconnect to landscape and nature, but also to provide some amount of food in times and circumstances where food cost and availability may be difficult. A city of gardeners, as with the experience of war time victory gardens, can produce a great deal and enhance urban resilience in the face of weather and energy shocks that may affect the flow and sourcing of food from outside a city [48]. Maintenance of economic viability of peri-urban farms and regional agriculture, as well as expanding opportunities to grow food in the leftover spaces of cities, from rooftops to re-claimed building sites, will strengthen the resilience of a city.

1.3.4 BIOPHILIC CITIES CAN HELP BUILD SOCIAL CAPITAL AND TRUST

Biophilic cities and city initiatives can also, in many important ways, help to bring people together in pursuit and enjoyment of common interests and concerns and can expand and strengthen social networks and capital. As Table 2 suggests, one way in which we can judge the biophilic *bone fides* of a city is by understanding the number of its citizens engaged in one or another activity centered around learning about or directly experiencing nature—watching birds, hiking and camping, participating in a variety of ecological and nature-oriented clubs and social activities. Outside activities can be solitary, to be sure, but they are often collective and social in nature and often result in significant friendships and social relationships.

Biophilic urbanism can also often result in the development of important new personal competencies and contribute in important ways to a sense of wellbeing and meaning. Social isolation is a significant and growing problem in northern cities and societies, and efforts centered on participation in nature—viewing and watching, interacting with, restoring and repairing—can help in significant ways to overcome that isolation.

Forests, parks and green spaces are important places for socializing and for forming new friendships. Seeland, Dubendorfer and Hansmann [39], in a study of Zurich, conclude that such green areas are important vehicles for social inclusion and for integration of immigrant youth. "Meeting and communicating in open spaces … can be a platform for breaking up social segregation, and therefore, public places are indispensable for meeting and establishing contacts" [39]. Similar conclusions have been reached about community gardens, and such green areas integrated into living environments often serve as ready spaces for breaking-down barriers of various sorts in urban neighborhoods and providing opportunities for informal contact and socialization.

The green spaces of a city offer important sites for residents to come together and to build a degree of cohesion and trust. Furthermore, nature in the city can have important civilizing functions. Kuo and Sullivan, for example, have found that reported crime rates are lower in public housing projects with greater levels of greenery [49]. In another study of public housing residents, Sullivan and Kuo found differences in reported levels of violence and aggression depending on the presence of nearby nature: "levels of aggression and violence were significantly lower", they report, "among individuals who had some nearby nature outside their apartments than among their counterparts who lived in barren conditions" [50].

Friendships and social interaction provided by direct participation in nature activities can in turn help to strengthen adaptive capacity. Evidence of the recuperative power of friendships is growing, suggesting that a key pathway to enhanced individual and family resilience is through programs and activities that foster socialization and help build extensive and deep friendships [51,52]. Research from psychology suggests that the presence of friends will allow individuals and families to be more resilient in many ways, helping reduce perceptions of the insurmountable nature of problems and contributing to a sense that life-challenges can be overcome [52].

Biophilic green systems in cities can create the context and opportunities for people to come together. In Zurich, for example, one of the few systematic efforts to daylight streams in that city (i.e., to bring them back to the surface, from underground pipes) has created new green spaces in neighborhoods and new opportunities for kids to explore and spend time outside. There, fifteen years of work repairing and daylighting streams in urban neighborhoods has resulted in 16,000 meters of restored streams and the completion of some forty stream daylighting projects [53]. Buchli and Conraden report on the social value and benefits of these efforts [53]:

> With respect to the *Cultural System*, the relationship between humans and nature in urban neighborhoods and the understanding of, and care in dealing with, nature are important positive factors, which are strengthened by the open streams. The *Social Systems* include increased social interaction and networking in the neighborhoods, citizen participation in the planning and implementation process and interdisciplinary cooperation across the city government departments. Many stream opening projects have contributed to strengthening the social systems and in turn have been improved by citizen participation.

Participation and engagement with nature in cities is a key aspect of a biophilic urbanism and can take many forms. These include participation in a variety of nature clubs and organizations and activities, from bird-watching to native plant societies to fungi and food foraging. Some are more recreational (hiking clubs), while others are more educational, but all entail a form of active engagement in the community. Citizens of a biophilic city care about and are intimately engaged in experiencing and learning about nature. They (ideally) exhibit curiosity and experience fascination and wonder at the wildness and majesty in and around cities. There are a variety of social and health benefits provided through such participation and engagement. New friendships are forged, socialization and interaction fostered and additional meaning added to one's life (hopefully). Involvement in these organizations and the activities undertaken is also simply fun and delivers many of the important positive mood and restorative qualities already mentioned.

We are not the first to identify these potential connections and relationships, and Paton, Kelly and Doherty [54] have written a thorough chapter examining "natural environment as a source of adaptive capacity", with a focus on natural disasters. They suggest, among other factors, the important role that direct participation in local environment can have. "That is, there is evidence that exposure [to environment] can engender a positive and protective attitude to the local environment (e.g., Park Care, Bush Care and Land Care Groups). This exposure may come from passive or active participation in natural settings or from living in or near peri-urban areas"

[54]. Many Australian cities have extensive urban Bush Care groups, a volunteer army of citizens who engage in environmental clean-up, vegetation-planting and restoration work. In the City of Brisbane alone, for example, there are more than 120 active urban bush care groups [28]. Evidence suggests then that through participation bonds with place and caring for both place and community will likely increase as a result. Paton, Kelly and Doherty summarize these potential relationships [54]:

> It is evident … that interaction with the environment can perform a restorative function and contribute to well-being. It can also act as a protective factor in regard to mitigating present and future stress, act as a catalyst for meaningful interaction that facilitates the development and/or maintenance of adaptive capacities (e.g., self and collective efficacy) and contribute to the development and maintenance of attachment and commitment to place/community.

Engaging citizenry in a host of hands-on nature activities, then, can be an effective strategy for enhancing adaptive capacity and resilience. There are a number of exemplary programs for engaging the public in active participation on nature in cities around the world. These include master naturalists programs, which further introduce the benefits of mastery of subject. One of the best the Texas Master Naturalist program, a partnership of the Texas Parks and Wildlife Department and Texas AgriLife Extension Service. Through the program citizens receive training in the nature and natural history of the state and if they meet the requisite hours of classroom, field study and volunteer service, can become certified master naturalists. Programs like this have been wildly popular and successful on almost every measure. Some 6000 Texans have gone through the training, providing more than 1 million hours of volunteer labor since the program began in 1998 [55]. A wide range of volunteer activities has occurred through the program, from wetlands restoration, to nesting sea turtle patrol, to conducting field surveys.

A biophilic city also places priority on educating children and young people about the nature around them and fostering these natural connections at an early stage. In Houston, Texas, for instance, a coalition of public and private entities called Houston Wilderness, has been actively educating children about the region's biodiversity and has produced a "wilderness passport" used by all fourth-grade classes in the Houston Independent School District. The passport is a guide to the region's ten ecosystems and encourages the visiting of all of them [56].

There is also growing recognition that citizens can play an essential role in helping collect data and helping in ecological sciences. This can happen through urban bio-blitzes, for instance, intensive 24-hour inventories of the biodiversity in a particular place, often within a city. Increasingly bio-blitzes, with significant citizen engagement, have occurred in a number of U.S. cities (including, for instance, one organized in Central Park, in New York City, that uncovered an entirely new species of centipede, *Nannarrup hoffmani*, a very small invertebrate, with 82 legs, as it turns out). Furthermore, cities can cultivate potentially life-long citizen scientists at an early stage, such as New York's effort to engage seventh graders in public and private schools throughout in monitoring salamanders at city parks [57].

City-based desert conservation efforts in the greater Phoenix area of the U.S. offer additional positive example, especially the efforts of citizens and a grassroots effort to create and manage a large desert preserve system in the City of Scottsdale. Here, amazing biophilic physical resources are available to a large segment of the population (in the form of extensive desert habitat and hiking trails, now covering more than 17,000 acres, in close proximity to neighborhoods), but are also complimented by opportunities to participate in the management of these spaces. The McDowell Sonoran Conservancy shares responsibility for operating and managing the desert preserve (along with the City of Scottsdale) and impressively has an army of 400 volunteers, who undertake minimum training and become "stewards," participating in a variety of ways in the management of, and education about, this immense reservoir of urban nature. Interestingly as budgets for parks and recreation departments all over Arizona (and the U.S.) decline, deep cuts in parks personnel have occurred. The Scottsdale model of citizen-based management reflects in many ways a more resilient approach to caring for these resources in times of severe budgetary limits.

Precisely what the long-term individual impacts of these kinds of biophilic engagement are remains to be seen and is an important area for future research. However, it stands to reason that such grass-roots involvement has many positive benefits, including a deep level of intimacy with environment and place and the cultivation of valuable friendships and social networks, among other benefits.

1.4 OBSTACLES TO BIOPHILIC RESILIENCE

The power and value of biophilic urbanism as an urban resilience and sustainability strategy is clear, and a number of cities around the world have developed and are implementing ambitious programs for restoring, protecting and expanding the nature in and around them. As this paper argues, the resilience benefits are many, with biophilic investments—from trees and urban forests to healthy wetland and riparian systems—serving to expand adaptive capacity in the face of a "perfect storm" of shocks and stressors that global cities will face in the decades ahead. These will include heat waves, drought and other likely effects associated with climate change, as well as natural disasters and a host of potential resource shocks and scarcities, such as decline in global oil supplies and availability of water and food.

In addition to the many and considerable direct benefits from biophilic cities, there are many ways in which access to nature will make individuals, families and communities healthier and happier and will help to forge new social connections and friendships, that should make such cities more resilient. Healthier, more socially-connected individuals, families and communities will increase the likelihood of successful adaptation to this dynamic future.

But there are challenges ahead and much work still to be done. There are considerable obstacles that remain in making cities more natural and nature-full. Some of these obstacles are social and cultural, while others are economic and legal. In many cities, for instance, many examples of biophilic urban design and urbanism are prohibited or made difficult by existing codes (e.g., prohibiting the disconnecting of residential downspouts and more localized and natural forms of storm water management). Greener, more natural urban environments may be constrained by cultural and aesthetic bias—failure to see the beauty in native vegetation, for instance, or in urban landscaping other than manicured and finely mowed turfgrass, and for many urbanites, there is actually a surprising amount of fear about nearby nature that must be overcome (i.e., spiders, bats and even coyotes!).

Some resistance to a more biophilic urban realm, then, may actually come from urbanites themselves. Interestingly, new experimental research suggests that individuals underestimate the benefits and enjoyment they will receive from spending time in nature [58]. Busy schedules, heavy work commitments and over dependence on indoor-lives and car-dependent lives represent further obstacles.

And there are obstacles presented by the prevailing short-term centered political and economic decision making mechanisms. Short term economic cost may be an impediment, for instance, in installation of green neighborhood or project features, such as green walls and green roofs, though the long term savings almost always dwarfs these short term costs (and thus a need to find creative ways to encourage long term and full-cost accounting).

There are also other important ways in which cities could be biophilic that have not been explored in this paper. For instance, one notion of a biophilic city is a city that learns from, mimics and is modeled after natural systems in its functioning. Increasingly, it is recognized that a city that strives for more local and regional sourcing of material inputs, and an urban metabolism that is more circular in nature, will be more sustainable and resilient. Cities might also be viewed as more biophilic if their buildings and built environments reflect the shapes and forms of nature, attributes that may deliver important emotional and social benefits.

As this paper has pointed out in numerous places, there are still many important research and policy questions to be answered in moving cities toward biophilic urbanism. The pathways identified above are largely tentative and require much additional research and practical demonstration. We have little understanding, for instance, of what the minimum level of nature, or access to nature, is that is required by urbanites to live healthy, happy and resilient lives (and the cumulative impacts of different green features and combinations of green features, in urban environments). Much additional research is needed to better understand, moreover, the full package of resilient benefits that is provided by a biophilic cities and even the different basic ways in which urban biophilia can manifest (which will likely vary by geography and climate).

REFERENCES

1. Wilson, E.O. Biophilia and the conservation ethic. In Biophilia: The Human Bond with Other Species; Kellert, S., Wilson, E.O., Eds.; Harvard University Press: Cambridge, MA, USA, 1993.
2. Wilson, E.O. Biophilia; Harvard University Press: Cambridge, MA, USA, 1984.
3. Wilson, E.O. The Creation: An Appeal to Save Life on Earth; Norton and Company: New York, NY, USA, 2007.
4. Leather, P.; Pyrgas, M.; Beale, D.; Lawrence, C. Windows in the workplace: Sunlight, views, and occupational stress. Environ. Behav. 1998, 30, 739–762.
5. Herschong, L.; Roger, W.; Stacia, O. Daylighting impacts on human performance in schools. J. Illum. Eng. Soc. Summer 2002, 31, 101–114.
6. Ulrich, R. View through a window may influence recovery from surgery. Science 1984, 224, 421.
7. DeVries, S.; Verheij, R.A.; Groenewegen, P.P.; Spreeuwenberg, P. Natural environments—healthy environments? An exploratory analysis of the relationship between greenspace and health. Environ. Plan. A 2003, 35, 1717–1731.
8. Nielsen, T.S.; Hansen, K.B. Do green areas affect health? Results from a danish survey on the use of green areas and health indicators. Health Place 2007, 13, 839–850.
9. MIND, Ecotherapy: The Green Agenda for Mental Health; MIND: London, UK, 2007.
10. Van den, B.; Agnes, E.; Hartig, T.; Straats, H. Preference for nature in urbanized societies: Stress, restoration and the pursuit of sustainability. J. Soc. Issues 2007, 63, 88–89.
11. Hartig, T.; Mang, M.; Evans, G.W. Restorative effects of natural environmental experience. Environ. Behav. 1991, 33, 3–26.
12. Mitchell, R.; Popham, F. Effect of exposure to natural environment on health inequalities: An observational population study. Lancet 2008, 373, 1655–1660.
13. Atchley, R.A.; Strayer, D.L.; Atchley, P. Creativity in the wild: Improving creative reasoning through immersion in natural settings. PLoS One 2012, 7, e51474.
14. Aspinall, P.; Panagiotis, M.; Richard, C.; Jenny, R. The urban brain: Analyzing outdoor physical activity with mobile EEG. Br. J. Sport Med. 2013.
15. Beatley, T. Biophilic Cities: Integrating Nature into Urban Design and Planning; Island Press: Washington, DC, USA, 2010.
16. Girling, C.; Ronald, K. Skinny Streets and Green Neighborhoods: Design for Environment and Community; Island Press: Washington, DC, USA, 2005.
17. Peter, N.; Kenworthy, J. Sustainability and Cities; Island Press: Washington, DC, USA, 1999.
18. Timothy, B.; Manning, K. The Ecology of Place; Island Press: Washington, DC, USA, 1997.
19. Peter, N.; Beatley, T.; Boyer, H. Resilient Cities: Responding to Peak Oil and Climate Change; Island Press: Washington, DC, USA, 2008.
20. Godscalk, D.R. Urban hazard mitgation: Creating resilient cities. Nat. Hazards Rev. 2003, 4, 136–143.

21. Timothy, B. Planning for Coastal Resilience: Best Practices for Calamitous Times; Island Press: Washington, DC, USA, 2009.
22. Walker, B.; Salt, D. Resilience Thinking: Sustaining Ecosystems and People in a Changing World; Island Press: Washington, DC, USA, 2006.
23. Costanza, R.; MItsch, W.J.; Day, J.W. A new vision for New Orleans and the Mississippi Delta: Applying ecological economics and ecological engineering. Front. Ecol. Environ. 2006, 4, 465–472.
24. Costanza, R.; Pérez-Maqueo, O.; Martinez, M.L.; Sutton, P.; Anderson, S.J.; Mulder, K. The value of coastal wetlands for hurricane protection. AMBIO J. Hum. Environ. 2008, 37, 241–248.
25. US Environmental Protection Agency. Filtration Avoidance. Available online: http://www.epa.gov/region2/water/nycshed/filtad.htm/ (accessed on 7 June 2013).
26. Denver Water, Forest-to-Faucet Partnership. Available online: http://www.denverwater.org/supplyplanning/watersupply/partnershipUSFS/ (accessed on 7 June 2013).
27. Ryerson University, 2005. Report on the Environmental Benefits and Costs of Green Roof Technology for the City of Toronto. Available online: http://www.toronto.ca/greenroofs/pdf/fullreport103105.pdf (accessed on 7 June 2013).
28. Beatley, T.; Newman, P. Green Urbanism Down Under; Island Press: Washington, DC, USA, 2008.
29. Hazard Resistant Landscaping. Charleston, South Carolina. Available online: http://www.charlestoncounty.org/printer/www/departments/BuildingInspections/ProjectImpact/landscaping_Brochure.pdf (accessed on 7 June 2013).
30. Sustainable Cities. Stuttgart: Cool City. Available online: http://www.dac.dk/en/dac-cities/sustainable-cities-2/all-cases/green-city/stuttgart-cool-city/ (accessed on 7 June 2013).
31. Wolf, K.L. City trees, nature and physical activity: A research review. Arborist News 2008, 17, 22–24.
32. Manzo, L.C. Towards a revisioning of emotional relationships with places. J. Environ. Psychol. 2003, 23, 47–61.
33. Florida, R. The Rise of the Creative Class; Basic Books: New York, NY, USA, 2003.
34. Manzo, L.C.; Perkins, D.D. Finding common ground: The importance of place attachment to community participation and planning. J. Plan. Lit. 2006, 20, 335–350.
35. Schantz, P.; Stigell, E. Are Green Elements Principal Pull Factors for Physical Activity; Swedish School for Sport and Health Sciences: Stockholm, Sweden, 2008.
36. Trust for Public Land. In Parks and Playground Use up in Down Economy; TPL: San Francisco, CA, USA, 2009.
37. Cobb, D.T. Reclaiming Our Food; Storey Press: Adams, MA, USA, 2011.
38. Louv, R. Last Child in the Woods: Saving our Children from Nature Deficit Disorder; Algonquin Press: Chapel Hill, NC, USA, 2005.
39. Klaus, S.; Dubendorfer, S.; Hansmann, R. Making friends in Zurich's urban forests and parks: The role of public green space for social inclusion of youths from different cultures. For. Policy Econ. 2009, 11, 10–17.
40. Wells, N.; Evans, G. Nearby nature: A buffer of life stress among rural children. Environ. Behav. 2003, 35, 311–330.

41. Kaiser Family Foundation, 2010. Generation M2: Media in the Lives of 8-to-18 Year Olds. Available online: http://kff.org/other/report/generation-m2-media-in-the-lives-of-8-to-18-year-olds/ (accessed on 7 June 2013).
42. City of New York. Family Camping. Available online: http://www.nycgovparks.org/programs/rangers/explorer-programs#family-camping/ (accessed on 7 June 2013).
43. Winterman, D. The Surprising Uses for Birdsong. Available online: http://www.bbc.co.uk/news/magazine-22298779/ (accessed on 7 June 2013).
44. Godby, G. Outdoor Recreation, Health, and Wellness: Understanding and Enhancing the Relationship. Resources for the Future Discussion Paper, May. Available online: http://www.rff.org/documents/RFF-DP-09-21.pdf (accessed on 7 June 2013).
45. Curtin, S. Wildlife tourism: The intangible, psychological benefits of human-wildlife encounters. Curr. Issues Tour. 2009, 12, 451–474.
46. Wolch, J. Zoopolis. Cap. Nat. Soc. 1996, 7, 21–47.
47. Weinstein, N.; Przybylski, A.K.; Ryan, R.M. Can nature make us more caring? Effects of immersionin nature on intrinsic aspirations and generosity? Person. Soc. Psychol. Bull. 2009, 35, 1315–1329.
48. Steel, C. Hungry City: How Food Shapes Our Lives; Random House: London, UK, 2009.
49. Kuo, F.E.; Sullivan, W.C. Environment and crime in the inner city: Does vegetation reduce crime? Environ. Behav. 2001, 33, 343–367.
50. Sullivan, W.; Kuo, F. Aggression and violence in the inner city: Effects of environment via mental fatigue. Environ. Behav. 2001, 33, 543–571.
51. Kroenke, C.H.; Kubzansky, L.D.; Schernhammer, E.S.; Holmes, M.D.; Kawachi, I. Social networks, social support, and survival after breast cancer diagnosis. J. Clin. Oncol. 2006, 24, 1105–1111.
52. Schnall, S.; Harber, K.D.; Stefanucci, J.K.; Proffitt, D.R. Social support and the perception of geographical slant. J. Exp. Soc. Psychol. 2008, 44, 1246–1255.
53. Buchli, R.; Conraden, F. The Zurich stream day-lighting program. In Enhancing Urban Environment by Environmental Upgrading and Restoration; Marsalek, J., Sztruhar, D., Urbonas, B., Eds.; Kluwer Academic Publishers: Dordrecht, The Netherlands, 2004.
54. Patton, D.; Kelly, G.; Doherty, M. Exploring the complexity of social and ecological resilience to hazards. In Disaster Resilience: An Integrated Approach; Patton, D., Johnston, D., Eds.; Charles C. Thomas: Springfield, IL, USA, 2006.
55. Texas Parks and Wildlife. Texas Master Naturalist. Available online: http://txmn.org/ (accessed on 7 June 2013).
56. Wilderness Passport: A Guide for Teachers and Scout Leaders; Houston Wildernes: Houston, TX, USA.
57. Foderado, L. Unleashing the Scientist in the Student. Available online: http://www.nytimes.com/2011/10/08/nyregion/salamander-study-enlists-new-york-city-seventh-graders.html?pagewanted=all&_r=0/ (accessed on 7 June 2013).
58. Nisbet, E.K.; Zelenski, J.M. Underestimating nearby nature: Affective forecasting errors obscure the happy path to sustainability. Psychol. Sci. 2011, 22, 1101–1106.

CHAPTER 2

Reconnecting Cities to the Biosphere: Stewardship of Green Infrastructure and Urban Ecosystem Services

ERIK ANDERSSON, STEPHAN BARTHEL, SARA BORGSTRÖM, JOHAN COLDING, THOMAS ELMQVIST, CARL FOLKE, AND ÅSA GREN

2.1 INTRODUCTION

The rate of urban growth is unprecedented. The Earth System has become urbanized in the sense that decisions by the majority of the human population now living in cities affect the resilience of the entire planet (Seto et al. 2011). Urban demand for ecosystem services is a major driver behind global environmental change but the choices people make are often disconnected from their environmental imprint in distant places (Folke et al. 1997; Grimm et al. 2008). Much of urban growth has been at the expense of the capacity of terrestrial and marine systems to generate and sustain essential ecosystem services (Foley et al. 2005) and is currently challenging biophysical planetary boundaries for the world as we know it (Rockström et al. 2009). There is an urgent need to reconnect people in urban areas to the biosphere (Folke et al. 2011).

Contemporary society, urban lifestyles, and changes, such as the decline of traditional land uses in the peri-urban landscape, have changed the way people in cities perceive and interact with the biosphere (Turner et al. 2004). The physical and mental distance between urban consumers and the ecosystems supporting them mask the ecological implications of choices made (Rees and Wackernagel 1996; Folke et al. 1997). Instead of oblivious consumers, cities need engaged stewards that can help redirect urbanization into a driver of positive change for humanity and the life-supporting systems that we depend upon. But how can people living in cities with urban lifestyles be reconnected to the biosphere? How do we ensure sustainable generation, management, and governance of ecosystem services for human well-being in cities, as well as ensure that cities contribute to incentives for better stewardship of distant landscapes and seascapes?

Though providing but a fraction of the ecosystem services consumed, urban landscapes represent key arenas for learning about the way humans interact with the environment and what sustainable ecosystem stewardship might entail (Miller 2005; Chapin et al. 2010). The focus of this paper is on lessons learnt for stewardship of ecosystem services within urban social–ecological systems (Berkes and Folke 1998). We draw on more than 15 years of empirical work within cities in relation to ecosystem service generation in urban landscapes, particularly regulating ecosystem services, and their stewardship with enabling institutions (e.g., property rights), social networks and involvement of local user groups and civil society in green area management and governance. Our work in the Stockholm urban landscape, Sweden, has helped reveal green areas and ecosystem services not previously perceived in urban planning and clarified mismatches between institutions, governance, and urban ecosystems for human well-being (Electronic Supplementary Material, Appendix S1). We emphasize that an urban social–ecological approach (Electronic Supplementary Material, Appendix S2) reduces the tension between conservation and city expansion and provides directions for shifting urbanization patterns toward sustainability. We recognize that most of our empirical work is from one particular city set in a certain context, but believe that our concluding propositions for urban resilience building can communicate with other cities and inspire theoretical discussions.

2.2 URBAN SOCIAL–ECOLOGICAL SYSTEMS

2.2.1 THE URBAN LANDSCAPE

Often, green space in urban areas can be remnants of a cultural landscape with biodiversity-rich habitats (Barthel et al. 2005). Many cities incorporate prime habitats that sometimes are rare in the larger region. For example, in regions where land-use intensification has led to loss of landscape diversity and habitats, such as ponds and non-cultivated elements, cities subjected to other drivers have become refuges for species associated with these habitats (Colding and Folke 2009). However, biodiversity and landscape heterogeneity in cities should not only be seen in relation to surrounding hinterlands. Urban landscapes have evolved under extremely complex influences of changing land uses and management practices, sustaining some habitats and fundamentally altering others. We need a detailed understanding of what "green" infrastructure really means in the urban context as well as how the values have come to be (Kinzig et al. 2005; Colding et al. 2006).

Cities are rife with "novel ecosystems" (Hobbs et al. 2006), which deserve to be acknowledged for the values they possess in terms of biodiversity and ecosystem services. Comprehensive analyses of urban green spaces have shown that land uses such as private and public gardens, cemeteries, old brown-fields, and golf courses may contribute significantly to ecosystem services provided by the urban landscape (Colding et al. 2006; Goddard et al. 2010). Incentives, interests, and ambitions among managers and stakeholders and the institutional framework set the stage for management of such spaces and their ecosystem services (Andersson et al. 2007). Over time, this close interaction between human actors, the social context in which they are embedded and the landscape may lead to biodiversity-rich systems maintained as much by human stewardship (Barthel et al. 2005, 2010), the protection of land by the state (Borgström

2009), civil society, and socioeconomic factors (Hope et al. 2003) as by ecological processes.

Urban landscape mosaics are often characterized by small land-use patches and high heterogeneity. It has been suggested that landscape structure becomes ecologically important only when a certain habitat drops below a threshold level coverage (Andrén 1994). This means that spatial structure becomes a key concern in cities, both as ecological networks and adjoining areas (Colding 2007; Andersson and Bodin 2009). Even if there are calls for more integrated landscape approaches in urban planning (Poiani et al. 2000), those commonly concern the large scale green structure and as a result leave out the potential and small scale patches within the built up areas (Colding et al. 2006). These integrated approaches also have to overcome the organization of urban policy that is characterized by a multitude of separate sectors and that fail to acknowledge the complexity of urban social–ecological systems (Runhaar et al. 2009). Issues relating to urban ecosystem services involve a wide range of actors seldom adding up to a comprehensive whole (Ernstson et al. 2010).

Furthermore, when addressing issues of biodiversity, both urban planning and nature conservation policies tend to focus on the establishment of set-asides using formal protection with strong focus on threatened species and their habitats. Such approaches risk reinforcing the land-use dichotomy of conservation versus exploitation and simply miss and exclude many ecologically important land uses, their ecosystem services and the local stewards engaged (Colding et al. 2006). The location of urban protected areas is often the result of intricate negotiations between ecological, economic, and social interests. In many cases, the politics of decision-making processes makes it more difficult to muster arguments for protection and ecological recognition of such areas that are make sure that sites attractive for urban real-estate developers (Ernstson et al. 2008; Borgström 2009).

2.2.2 ECOSYSTEM SERVICE IN URBAN AREAS

Green infrastructure in cities generates a diversity of ecosystem services (Jansson and Nohrstedt 2001). While we begin to understand the importance of urban green areas we still have a limited understanding of the mechanisms behind the generation of urban ecosystem services. The most commonly articulated link between urban green space and human well-being in current urban planning is through so called cultural services, e.g., recreation and health (Tzoulas et al. 2007). Also provisioning services, like food production in, for example, home gardens (Altieri et al. 1999; Krasny and Tidball 2009) and links to biodiversity conservation have been in focus (Goddard et al. 2010; van Heezik et al. 2012).

The studies reported here focus on the link to human well-being through regulating ecosystem services, such as seed dispersal, pest regulation, and pollination. These services are generated by complex interactions in urban social–ecological systems, and not by ecosystems alone (Andersson et al. 2007) as human activities may both promote service providers (Kremen 2005) and make services available to the beneficiaries (Fig. 1). This serves to illustrate the connection between biodiversity and ecosystem services (Kremen 2005) and the role of biodiversity for social–ecological resilience also in urban areas.

Many ecosystem services need to be locally provided in urban landscapes for easy access and use by a greater set of city-inhabitants, e.g., daily nature encounters, noise reduction, absorption of pollutants in water and air. The small size of many urban land-use patches make it difficult if not impossible to promote the generation of the full range of desired ecosystem services within individual patches. A closer investigation of regulating services reveals spatial and temporal interdependencies seldom recognized by governance structures. Many regulating services, including seed dispersal, pest regulation, and pollination, are not restricted to the areas where they originate but transcend habitat boundaries and affect also the surrounding landscape (Jansson and Polasky 2010; Blitzer et al. 2012). Such services may depend on functional connectivity (Fahrig et al. 2011) between different habitats, implying that a landscape perspective on management and planning for urban ecosystem services is often necessary (Colding 2007; Ernstson et al. 2010).

For example, Lundberg et al. (2008) showed how the preservation of a highly valued recreational oak-dominated landscape benefits from seed dispersing birds that also need coniferous forest. The coniferous forests tend to be located outside the recreational landscape and separated from it by administrative boundaries. Jansson and Polasky (2010) quantified the

Figure 1. Local user groups and stewardship of regulating ecosystem services in urban green areas. a Domestic gardens support biodiversity and species of significance in, e.g., pest control and seed dispersal (photo Carl Folke). b Allotment gardens provide critical habitats and food sources during vulnerable animal life history stages (photo Stephan Barthel). c Community gardens generate ecosystem services like pollination that spill over into the wider landscape (photo Johan Colding). d Urban golf courses function as stepping stones for keystone species with ponds hosting amphibians including endangered and keystone species (photo Stefan Lundberg). (e) Trees improve air quality and sequester carbon (photo Azote). f Green spaces within cities consist of remnants of biodiversity-rich cultural habitats in an otherwise fragmented landscape (photo Jakob Lundberg)

change in an ecosystem service over time and demonstrated how temporal dynamics may unintentionally erode the capacity to grow alternative crops in an agricultural system. Non-cultivated lands together with rape fields could sustain pollination and pollinator diversity, but were insufficient in themselves to maintain all pollinator species during periods of cereal production. The study showed how failure to address such dynamics eroded social–ecological resilience. By losing some of the pollinator species, the potential for response diversity diminished, making the regulating service more vulnerable to disturbance and change (Elmqvist et al. 2003).

The appreciation of green infrastructures in cities is often manifested in higher house prices close to green areas (Wittemyer et al. 2008). But appreciation and use as they are expressed today raise concerns about the long-term generation of ecosystem services and in particular regulating ecosystem services. For example, when green areas attract adjoining urban development they risk becoming isolated and thereby losing some of the biodiversity and related services that made them attractive in the first place (Borgström et al. 2012). Furthermore, high human population density and limited space in cities often result in demand for multifunctionality of green space, where stewardship of ecosystem services is confronted with multiple objectives, meanings, and conflicting interests (Borgström 2009; Ernstson and Sörlin 2009).

2.2.3 THE FORMATION OF STEWARDSHIP OF URBAN ECOSYSTEM SERVICES

Increasing people's awareness of how their actions impact the biosphere is not just a matter of close proximity to green areas, stewardship is about getting involved, which in turn may be facilitated by institutional designs and social movements. Today's institutions poorly match current changes in urban ecosystems (see Fig. 2; Borgström et al. 2006). Prospects for governance of urban ecosystem services, which strongly benefit from local stakeholder involvement, are becoming further limited when property rights systems change due to urbanization. Property right dynamics shaping human relationships to land can be quite influential, e.g., by helping counteract the growing disconnection of urban residents from nature (Pyle 1978). However, property right arrangements for the green infrastructure that produce urban ecosystem services seldom receive attention in urban settings in competition with other land uses. The global trend of privatization of public land in cities (Lee and Webster 2006) restricts people's ability to practically engage with urban ecosystems and their services, and if associated with loss of diversity this development might constrain the capacity to deal with change in effective ways. Common property systems, by which groups or a community of resource users share a common interest in resource management (Ostrom 1990), are rare in relation to urban ecosystems. This further reduces the opportunity for people and groups in cities to have meaningful interaction and provide stewardship of their local ecosystems (Andersson et al. 2007; Colding and Barthel 2013).

Diverse and new forms of property rights arrangements hold potential to play a much greater role in stewardship of urban landscapes than has hitherto been recognized (Colding and Barthel 2013). Furthermore, institutional diversity may not only increase diversity of land management approaches (Andersson et al. 2007), but also enhance self-organization of urban systems to adaptively deal with change, i.e., their social–ecological resilience (Folke et al. 2003). As institutional research suggests, having a multitude of property rights regimes that fit the cultural, economic, and geographic context in which they are to function (Hanna et al. 1996) appears also to be critical for resilience building of cities (Colding and Barthel 2013).

An important motivation for civic groups, at least the more affluent, to engage in stewardship of ecosystems in urban landscapes is sense-of-place, memory, and meaning (Andersson et al. 2007; Barthel et al. 2010). Social–ecological memory encapsulates the means by which knowledge, experience, and practice of ecosystem stewardship are captured, stored, revived, and transmitted through time (Barthel et al. 2010). For instance, in collectively managed gardens, community engagement results in a shared history manifested in artifacts, locally adapted organisms, trees, landscape features, and written accounts (Nazarea 2006; Barthel et al. 2010). These objects tend to outlive the practices that first shaped them and function as shared memory carriers between people and across generations (Barthel et al. 2010). Different

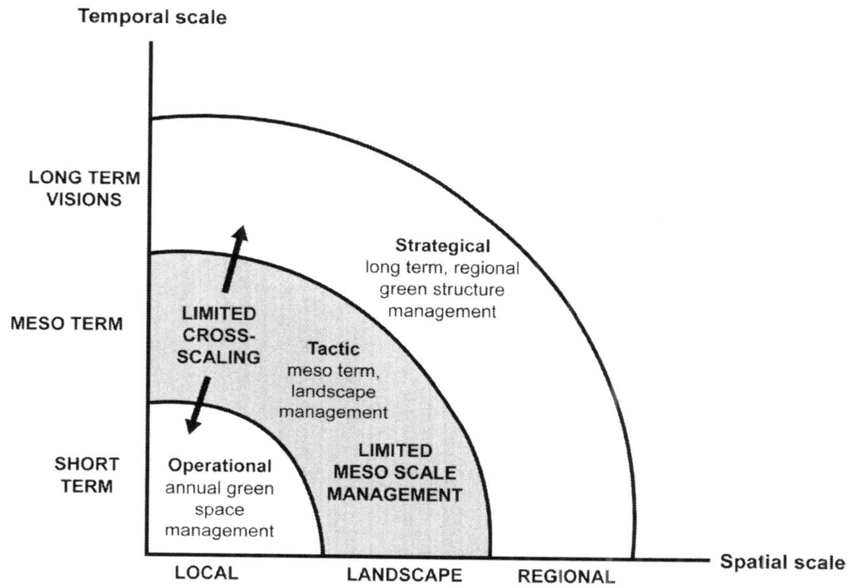

Figure 2. Comparatively little attention is paid to the meso-scale and cross-scale interactions are not recognized among planners and managers of urban green infrastructure (from Borgstrum et al. 2006).

forms of participation also carry shared memories, such as exchange of seeds for planting and oral traditions, which in combination with physical objects guide a portfolio of practices for how to deal with a changing social–ecological context, and local responses to such fluctuations. For instance, in some garden communities, a small percentage of 1 year's harvest is often saved for the next planting. Over time, this enhances the probability of locally adapted varieties of crops coevolved with human practices and local environmental conditions. Social–ecological memory in collectively managed gardens, for example, is favorable for the conservation of ecosystem service providers (Kremen 2005) normally associated with rural landscapes.

Current urban green spaces tend primarily to be managed at the local scale, where within-site qualitative characteristics are the focus

(Borgström et al. 2006; Andersson et al. 2007; Ernstson et al. 2010). However, the spatial and temporal dynamics of ecosystem services often demand co-operation and co-ordination across the landscape and administrative boundaries. Also, the full potential complexity of local engagement becomes evident first at an aggregate level (Fig. 3). User groups interact and form social networks whose structures may both facilitate and constrain collective action towards ecosystem management and stewardship (Ernstson et al. 2008, 2010). The formation of co-management is channeled through the ability of civil society organization to build alliances between each other, and to government departments. It has been found that there are often more contacts between managers handling the same kind of area (e.g., cemeteries) than between neighboring green space managers, implying a neglect of plausible spatial ecological connections (Borgström et al. 2006; Ernstson et al. 2010). Actors able to connect over these boundaries, called brokers, are crucial as they greatly increase the opportunities for a diversity of actor groups to meet and exchange experiences. As historical (Walker 2007) and social movement research has indicated (Ansell 2003; Ernstson et al. 2008), urban green areas attracting a high diversity of interest and user groups seem to have higher chances of being protected and creating a social environment that nurture stewardship of ecosystem services because of increased potential for effective collective action and combination of knowledge and skills.

2.3 CONCLUSION AND IMPLICATIONS FOR RECONNECTING URBAN AREAS TO THE BIOSPHERE

The understanding of how urban ecosystems work, how they change, and what limits their performance, can add to the understanding of ecosystem change and governance in general in an ever more human-dominated world with implications for Earth Stewardship (Chapin et al. 2010). The high concentration of people, the diverse preferences that individuals, groups, business, and the state have for the city and the various demands for ecosystem services will cause continuous tension, which urban planning systems should be set up to handle. The

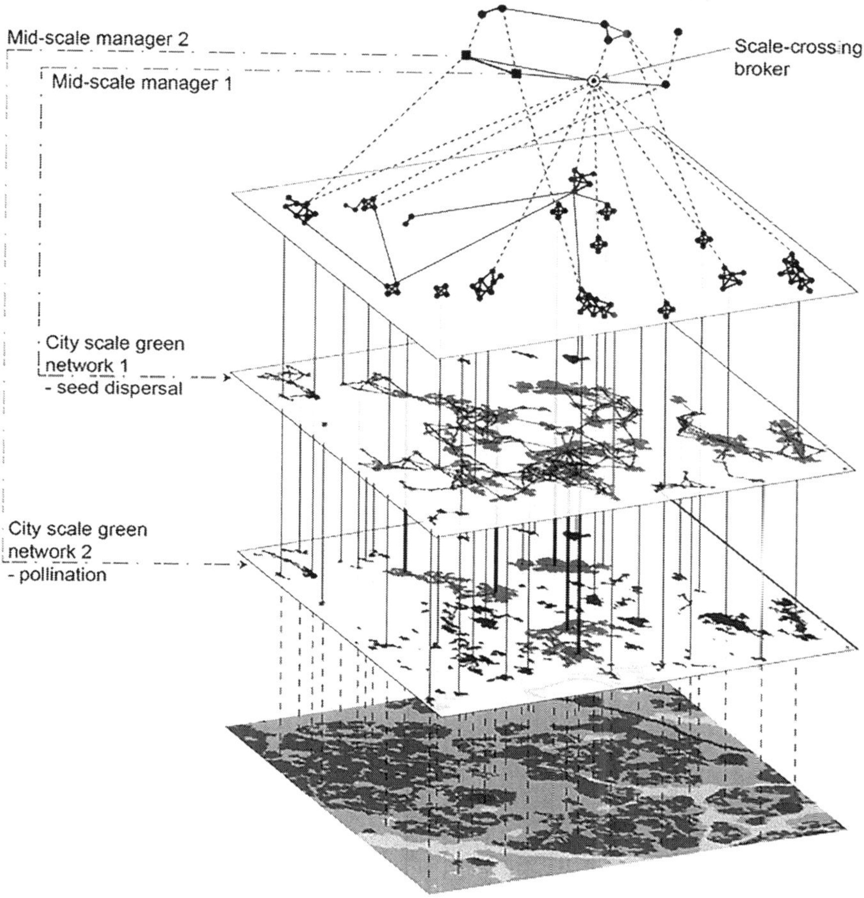

Figure 3. To support the continuous generation of urban ecosystem services, governance structures are needed that connect local experiential knowledge of ecosystem management with those of higher scale understanding outlined in the figure. In such arrangements, the broker position in social networks should be identified and strengthened since it may be needed to link ecosystem stewards across scales, and on different sides of sectoral and administrative boarders. Such scale-crossing brokers might be complemented with more ecologically focused mid-scale managers (Ernstson et al. 2010) (figure from Ernstson et al. 2010).

importance and potential of urban planning also highlight the need for more research in the resource scarce cities of the Global south where the role played by planning is smaller.

Biodiversity and ecosystem services in urban landscapes are generated by complex interactions between ecological processes and human activities and organization. In an attempt to tackle this complexity, the scope of the research reported here includes social science in combination with systems ecology, ranging from local ecological knowledge as a strong connector between people and their environment to issues like learning, social memory, property rights, social movements, social justice, and cultural narratives. We have also highlighted the importance of including governance perspectives of legal protection, of actors in civil society, of brokers, and of environmental movements in the co-production of ecosystem services and biodiversity and the challenges of overcoming mismatches between the social and ecological systems both in space and time. This broad social–ecological approach on urban ecology has yielded a number of findings that should be of interest for this booming field of research:

1. Current urban planning strategies often fail to acknowledge ecological and social synergies. Distinct social–ecological dynamics in seemingly similar patches result in quite different and potentially complementing profiles of biodiversity and ecosystem services that might be lost if this stewardship is not understood or nurtured.
2. Small-scale land-use heterogeneity makes spatial organization especially important. The length and diversity of borders, biophysical as well as administrative, call for careful thinking to make sure adjacency effects are positive and that regulating ecosystem services reach across the urban landscapes.
3. Cities hold unexplored potential for new urban spatial designs that integrate ecosystem services in the built environment, for restoring degraded ecosystems and for strengthening ecosystem functions through complementary designs of land use and urban green structures.
4. Mismatches between social and ecological boundaries are prevalent. The meso-scale between local and regional is insufficiently addressed. Governance structures are needed that connect local experiential knowledge of ecosystem management with those of higher scale understanding. In such arrangements, the broker

position in social networks should be identified and strengthened since it may be needed to link ecosystem stewards across scales, and on different sides of sectoral and administrative borders.

5. Different property rights lead to differences in practices, willingness to invest and learn about the system. Short-term tenure is more flexible while long term may lead to in-depth, situated knowledge and investment in restoration.
6. Participatory management approaches are critical for harnessing the diversity found within cities. These draw on diversity in the skill-bases that people and groups possess and also have the potential to provide more effective urban ecosystem management by taking into account multiple ways of knowing and evaluating urban land.

Cities could become laboratories where management strategies and governance structures for ecosystem stewardship are tested and evaluated. As most cities are not directly dependent on having all (especially provisioning) ecosystem services generated within-city boundaries, they are comparatively safe spaces for testing new governance structures and management practices within the domains where urban planning and design operate. For example, cities might be the best places to seek the answer to how diverse and contested interests in combination with limited space might be navigated to establish multifunctional land uses, an issue that will become increasingly important in many different social–ecological systems.

Cities arguably need to reduce their ecological footprint, but perhaps more importantly the character of the footprint need to change. A crucial step is to provide within-city opportunities for responsible stewardship to help reconnect citizens to the biosphere. In general, the promotion of "cognitive resilience building" for ecosystem stewardship in urban areas is central (Colding and Barthel 2013). It implies the perceptions, memory, and reasoning that people acquire from frequent interactions with local ecosystems, shaping peoples' experiences, world views, and values toward local ecosystems and ultimately toward the biosphere. To achieve institutional changes, further studies are needed to explore the wider political processes that shape and promote how biophysical processes become articulated as of value, for example through the use of framings like ecosystem services. In a world where soon two-thirds of the population will live in cities both the individual and institutional level of analysis is of fundamental importance. Together with further research on the ecological underpinnings of ecosystem services, not least the cultural, future long-term urban social–ecological research must deepen our understanding of whether and how local stewardship and engagement in practical management of green infrastructures increase biodiversity and availability of ecosystem services in metropolitan landscapes, and if and how it actually stimulates a wider awareness and articulation of our global reliance on ecosystem services and results in an urban footprint both smaller and less detrimental to the resilience of the biosphere.

REFERENCES

1. Altieri, M.A., N. Companioni, K. Cañizares, C. Murphy, P. Rosset, M. Bourque, and C.I. Nicholls. 1999. The greening of the "barrios": Urban agriculture for food security in Cuba. Agriculture and Human Values 16: 131–140.
2. Andersson, E., and Ö. Bodin. 2009. Practical tool for landscape planning? An empirical investigation of network based models of habitat fragmentation. Ecography 32: 123–132.
3. Andersson, E., S. Barthel, and K. Ahrné. 2007. Measuring social–ecological dynamics behind the generation of ecosystem services. Ecological Applications 17: 1267–1278.
4. Andrén, H. 1994. Effects of habitat fragmentation on birds and mammals in landscapes with different proportions of suitable habitat—A review. Oikos 71: 355–366.
5. Ansell, C.K. 2003. Community embeddedness and collaborative governance in the San Francisco Bay Area Environmental Movement. In Social movements and networks—Relational approaches to collective action, ed. M. Diani, and D. McAdam, 123–144. Oxford, UK: Oxford University Press.
6. Barthel, S., J. Colding, T. Elmqvist, and C. Folke. 2005. History and local management of a biodiversity-rich, urban, cultural landscape. Ecology and Society 10: 10.
7. Barthel, S., C. Folke, and J. Colding. 2010. Social–ecological memory in gardening: Retaining the capacity for management of ecosystem services. Global Environmental Change 20: 255–265.

8. Berkes, F., and C. Folke. 1998. Linking social and ecological systems: Management practices and social mechanisms for building resilience. Cambridge, UK: Cambridge University Press.
9. Blitzer, E.J., C.F. Dormann, A. Holzschuh, A.-M. Kleind, T.A. Rand, and T. Tscharntke. 2012. Spillover of functionally important organisms between managed and natural habitats. Agriculture, Ecosystems and Environment 146: 34–43.
10. Borgström, S.T. 2009. Patterns and challenges of urban nature conservation—A study of southern Sweden. Environment and Planning A 41: 2671–2685.
11. Borgström, S.T., T. Elmqvist, P. Angelstam, and C. Alfsen-Norodom. 2006. Scale mismatches in management of urban landscapes. Ecology and Society 11: 16.
12. Borgström, S., S.A.O. Cousins, and R. Lindborg. 2012. Outside the boundary—Land use changes in the surroundings of urban nature reserves. Applied Geography 32: 350–359.
13. Chapin, F.S., S.R. Carpenter, G.P. Kofinas, C. Folke, N. Abel, W.C. Clark, P. Olsson, D.M. Stafford Smith, et al. 2010. Ecosystem stewardship: Sustainability strategies for a rapidly changing planet. Trends in Ecology & Evolution 25: 241–249.
14. Colding, J. 2007. "Ecological land-use complementation" for building resilience in urban ecosystems. Landscape and Urban Planning 81: 46–55.
15. Colding, J., and S. Barthel. 2013. The potential of "Urban Green Commons" in the resilience building of cities. Ecological Economics 86: 156–166.
16. Colding, J., and C. Folke. 2009. The role of golf courses in biodiversity conservation and ecosystem management. Ecosystems 12: 191–206.
17. Colding, J., J. Lundberg, and C. Folke. 2006. Incorporating green-area user groups in urban ecosystem management. AMBIO 35: 237–244.
18. Elmqvist, T., C. Folke, M. Nyström, G. Peterson, J. Bengtsson, B. Walker, and J. Norberg. 2003. Response diversity, ecosystem change, and resilience. Frontiers in Ecology and the Environment 1: 488–494.
19. Ernstson, H., and S. Sörlin. 2009. Weaving protective stories: Connective practices to articulate holistic values in the Stockholm National Urban Park. Environment and Planning A 41: 1460–1479.
20. Ernstson, H., S. Sörlin, and T. Elmqvist. 2008. Social movements and ecosystem services—The role of social network structure in protecting and managing urban green areas in Stockholm. Ecology and Society 13: 39.
21. Ernstson, H., S. Barthel, E. Andersson, and S.T. Borgström. 2010. Scale-crossing brokers and network governance of urban ecosystem services: The case of Stockholm, Sweden. Ecology and Society 15: 28.
22. Fahrig, L., J. Baudry, L. Brotons, F.G. Burel, T.O. Crist, R.J. Fuller, C. Sirami, G.M. Siriwardena, et al. 2011. Functional landscape heterogeneity and animal biodiversity in agricultural landscapes. Ecology Letters 14: 101–122.
23. Foley, J.A., R. DeFries, G.P. Asner, C. Barford, G. Bonan, S.R. Carpenter, F.S. Chapin, M.T. Coe, et al. 2005. Global consequences of land use. Science 309: 570–574.
24. Folke, C., A. Jansson, J. Larsson, and R. Costanza. 1997. Ecosystem appropriation by cities. AMBIO 26: 167–172.
25. Folke, C., J. Colding, and F. Berkes. 2003. Building resilience and adaptive capacity in social–ecological systems. In Navigating social–ecological systems:

Building resilience for complexity and change, ed. C. Folke, F. Berkes, and J. Colding, 352–387. Cambridge, UK: Cambridge University Press.

26. Folke, C., Å. Jansson, J. Rockström, P. Olsson, S.R. Carpenter, F.S. Chapin, A.-S. Crépin, G. Daily, et al. 2011. Reconnecting to the biosphere. AMBIO 40: 719–738.
27. Goddard, M.A., A.J. Dougill, and T.G. Benton. 2010. Scaling up from gardens: Biodiversity conservation in urban environments. Trends in Ecology & Evolution 25: 90–98.
28. Grimm, N.B., S.H. Faeth, N.E. Golubiewski, C.L. Redman, J. Wu, X. Bai, and J.M. Briggs. 2008. Global change and the ecology of cities. Science 319: 756–760.
29. Hanna, S., C. Folke, and K.-G. Mäler. 1996. Rights to nature: Ecological, economic, cultural, and political principles of institutions for the environment. Washington, DC: Island Press.
30. Hobbs, R.J., S. Arico, J. Aronson, J.S. Baron, P. Bridgewater, V.A. Cramer, P.R. Epstein, J.J. Ewel, et al. 2006. Novel ecosystems: Theoretical and management aspects of the new ecological world order. Global Ecology and Biogeography 15: 1–7.
31. Hope, D., C. Gries, W.X. Zhu, W.F. Fagan, C.L. Redman, N.B. Grimm, A.L. Nelson, C. Martin, et al. 2003. Socioeconomics drive urban plant diversity. Proceedings of the National Academy of Sciences of the United States of America 100: 8788–8792.
32. Jansson, Å., and P. Nohrstedt. 2001. Carbon sinks and human freshwater dependence in Stockholm County. Ecological Economics 39: 361–370.
33. Jansson, Å., and S. Polasky. 2010. Quantifying biodiversity for building resilience for food security in urban areas: Getting down to business. Ecology and Society 15: 20.
34. Kinzig, A.P., P. Warren, C. Martin, D. Hope, and M. Katti. 2005. The effects of human socioeconomic status and cultural characteristics on urban patterns of biodiversity. Ecology and Society 10: 23.
35. Krasny, M., and K. Tidball. 2009. Community gardens as contexts for science, stewardship, and civic action learning. Cities and the Environment 2: 8.
36. Kremen, C. 2005. Managing ecosystem services: What do we need to know about their ecology? Ecology Letters 8: 468–479.
37. Lee, S., and C. Webster. 2006. Enclosure of the urban commons. GeoJournal 66: 27–42.
38. Lundberg, J., E. Andersson, G. Cleary, and T. Elmqvist. 2008. Linkages beyond borders: Targeting spatial processes in fragmented urban landscapes. Landscape Ecology 23: 717–726.
39. Miller, J.R. 2005. Biodiversity conservation and the extinction of experience. Trends in Ecology & Evolution 20: 430–434.
40. Nazarea, D.V. 2006. Local knowledge and memory in biodiversity conservation. Annual Review of Anthropology 35: 317–335.
41. Ostrom, E. 1990. Governing the commons: The evolution of institutions for collective action, 280. New York: Cambridge University Press.
42. Poiani, K.A., B.D. Richter, M.G. Anderson, and H.E. Richter. 2000. Biodiversity conservation at multiple scales: Functional sites, landscapes, and networks. BioScience 50: 133–146.

43. Pyle, R.M. 1978. The extinction of experience. Horticulture 56: 64–67.
44. Rees, W.E., and M. Wackernagel. 1996. Urban ecological footprints: Why cities cannot be sustainable—And why they are a key to sustainability. In Our ecological footprint, reducing human impact on the earth, ed. M. Wackernagel, and W.E. Rees, 223–248. Gabriola Island: New Society Publishers.
45. Rockström, J., W. Steffen, K. Noone, Å. Persson, F.S. Chapin, E.F. Lambin, T.M. Lenton, M. Scheffer, et al. 2009. A safe operating space for humanity. Nature 461: 472–475.
46. Runhaar, H.A.C., P.P.J. Driessen, and L. Soer. 2009. Scientific commons: Sustainable urban development and the challenge of policy integration: An assessment of planning tools for integrating spatial and environmental planning in the Netherlands. Environment and Planning B: Planning and Design 36: 417–431.
47. Seto, K.C., M. Fragkias, B. Güneralp, and M.K. Reilly. 2011. A meta-analysis of global urban land expansion. PLoS ONE 6: e23777.
48. Turner, W.R., T. Nakamura, and M. Dinetti. 2004. Global urbanization and the separation of humans from nature. BioScience 54: 585–590.
49. Tzoulas, K., K. Korpela, S. Venn, V. Yli-Pelkonen, A. Kazmierczak, J. Niemelä, and P. James. 2007. Promoting ecosystem and human health in urban areas using green infrastructure: A literature review. Landscape and Urban Planning 8: 167–178.
50. van Heezik, Y.M., K.J.M. Dickinson, and C. Freeman. 2012. Closing the gap: Communicating to change, gardening practices in support of native biodiversity in urban private gardens. Ecology and Society 17: 34.
51. Walker, C. 2007. Redistributive land reform: For what and for whom? In The land question in South Africa: The challenge of transformation and redistribution, ed. L. Ntsebeza, and R. Hall, 132–151. Cape Town: HSRC Press.
52. Wittemyer, G., P. Elsen, W.T. Bean, A. Coleman, O. Burton, and J.S. Brashares. 2008. Accelerated human population growth at protected areas edges. Science 321: 123–126.

PART II

ADDRESSING CLIMATE CHANGE

CHAPTER 3

Implementing Local Climate Change Adaptation and Mitigation Actions: The Role of Various Policy Instruments in a Multi-Level Governance Context

E. CARINA H. KESKITALO, SIRKKU JUHOLA, NINA BARON, HÅKON FYHN, AND JOHANNES KLEIN

3.1 CHALLENGES OF CLIMATE POLICY

While increasing focus is placed on policy development in relation to climate change, it is yet unclear as to how successful or efficient these efforts developed so far are in terms of tackling climate change, both mitigation and adaptation. In reviewing the progress of adaptation, the latest report of the Intergovernmental Panel on Climate Change notes that implementation has so far been relatively limited [1] and discusses multiple potential limitations and barriers to this. With regard to mitigation, the need for further development within international commitments, as well as national and sub-national development, is highlighted. To a large extent, then, the social complexity of implementation and gaining political will for the development in the face of multiple stressors may be regarded as limiting progress towards managing climate change [2,3,4]. This limitation means that there is a need to problematize the structure of implementation further.

Climate policy, both mitigation and adaptation, by nature addresses a collective action problem, sometimes with no clearly identifiable organization or individual responsible to be steered to solve the problem. Within mitigation, the reduction of greenhouse gas emissions requires the participation, voluntary or involuntary, of all sectors of society from energy production to household energy consumption and from agricultural production to individual consumption habits. Within adaptation, the impacts of climate change are likely to affect all sectors of society and the natural world. The way in which this climate risk is realized depends on how individuals, societal sectors or ecosystems are exposed to hazards and how vulnerable they are.

Both adaptation and mitigation have emerged into an arena of environmental regulation, where related issues, i.e., energy policy, have been addressed in the last three decades with increasing numbers of measures. This leads to a situation where consideration needs to be given to what types of instruments are suitable and not conficting and what kinds of issues come up when these are applied jointly [5]. Thus, policy coordination with other policy arenas becomes crucial. The increasing focus on multiple ways of steering, ranging from regulative instruments to voluntary ones, opens up the debate whether there is a need for stronger regulation in terms of climate policy or whether non-mandatory measures are sufficient.

Recent research has shown that multi-sectoral strategies for both mitigation and adaptation in the EU context have been rather unsuccessful [6], potentially due to the complexity of integrating these into the existing policy environment. Studies have also shown that integration within similar agendas to that of climate policy, i.e., sustainability measures, the use of voluntary measures, has been fraught with difficulty [7,8,9,10]. Issues such as project dependency, cost and integration into existing programmes have hampered progress. This has meant that implementation has mainly been short-term and piecemeal alongside normal day-to-day planning [11].

This note discusses the issue of integration of climate change with the help of reviewing and comparing the regulative environment in three Nordic cases: adaptation to climate change at the municipal level in Denmark and Finland, and a case of local level mitigation in Norway. The term "regulation" in this paper is used to apply to the regulative environment at large, and thus covers the same area as that to which policy instruments (see Section 2.1) can be applied [12,13,14]. Our understanding of regulation thus includes both stronger regulation through legal and regulatory instruments, economic and financial instruments, and soft measures These three cases illustrate different patterns of implementation but share the feature that stronger overall regulation, through legal and regulatory instruments, is at present relatively limited. Hence, the cases provide a starting point for a critical inquiry into whether the current level of regulation is sufficient. In other words, if climate change policy is indeed a crucial consideration in public policy, is it enough that it (in particular on the adaptation side) has mainly been instituted through non-binding measures?

3.2 IMPLEMENTATION OF CLIMATE POLICY: MIXING OF OLD WITH NEW

The development of decision-making has been expressed in political science literature as not only that of government—i.e., steering by the state—but of governance, or steering amongst broader groups including not only state but also business and non-governmental organizations [15]. Multi-level governance has been defined as the negotiated exchange between institutions on multiple levels, including the local, regional and national organization of the state and other actors (ibid.). However, studies also with regard to areas other than the environment have noted that, while multi-level governance "enabled non-hierarchical linkages for interdependent policy coordination, it appeared scholars had underestimated the conditions needed to ensure effective policy co-ordination and, hence, convergence" [16].

Limitations have, among others, included "vertical co-ordination problems, including increasing transaction costs, which rose with 'the number of administrative levels and degree of subnational autonomy'" [16,17]. The multi-level setting also adds complexity in that not only state but other means of organization are recognized: "the 'geographical imaginary' of environmental politics, where discrete local, national and international arenas operated in parallel, needed to give way to an account which recognized the complex vertical linkages between state institutions and the emergence of new political spaces which exceed this lexicon" [18].

In particular, the EU structure has come to add to governance arrangements by resulting in new issue arrangements but also reconfiguring the state structure, for instance, through EU legislation, such as directives that are required to be enforced at state level. The formal structure of public policy and the instruments used are thus shifting from a purely state configuration to one that also includes the EU. As a result, we should also expect an increasing role of regulation and formal public policy instruments [19,20], despite much literature having focused on the increasingly networked nature of decision-making.

In addition to issues related to vertical coordination, it is also interesting to note that existing EU-level regulation can also significantly impact climate policy, whilst not directly addressing climate change concerns. A good example of this is the way in which the relatively far developed flood risk issue is dealt with in relation to an emerging policy issue of adaptation. This follows a pattern identified for emerging policy areas and also indicates that the largest regulative requirements for adaptation are emerging from other, better institutionalized policy areas [21,22]. In this, improvements in adaptation may largely result from a "download Europeanization," where cities revise their practices as a result of the required implementation of EU programs [20]. Much literature, however, has noted that, while these types of new arrangements are increasingly widespread, they supplement rather than replace the state structure [15,23]. The result is more complex as well, as binding regulation at EU level (directives) and national level are supplemented with more mixed and more increasingly "soft", i.e., non-mandatory, instruments as well as with market-based, or other non-state measures.

3.2.1 EMERGING AREAS, CHOICE OF POLICY INSTRUMENTS AND COMPLIANCE

While the choice of public policy instruments to regulate an area is often unproblematized, it has been argued that "instrumentation is really a political issue, as the choice of instrument—which, moreover, may form the object of political conflict—will partly structure the process and its results" [24]. Generally, public policy instruments are differentiated into three major categories: (1) legal and regulatory instruments; (2) economic and financial instruments; (3) "soft" instruments; in other words, what could be conceived of as the "sticks", the "carrots" and the "sermons" of public policy [21,24]. Some also add a fourth type of instrument to the list above, highlighting the role of institutional instruments that set the ground rules for interaction: for instance, property regimes [25,26,27]. In addition, soft instruments are often supported by non-government centered, market-based or other forms of regulation e.g., through development of standards or codes of conduct. These may be developed in coordination or with pressure from the state, or e.g., in relation to the market (see [14] for a conceptualization of decentred regulation). Whilst these may be of crucial importance, our focus here is mainly on public policy by the state: that is, on how the state institutionalizes issues with high political agreement such as those of adaptation and mitigation. We thus focus our discussion on the three first categories.

The first of these, legal and regulatory instruments or binding regulation, require "actors to act within some clearly defined boundaries of what is allowed and what is not allowed" [21]. They are thus usually reserved for norms with the highest level of agreement: it takes time, for instance, for a new policy area to reach such a stage that it is given a mandatory nature [21,22]. While implementation may still be limited, for instance in cases where laws are not enforced, these instruments are typically backed by threat of sanctions, such as fines or withdrawal of rights that also support some level of implementation. Legal and regulatory instruments thus both play a normative role in supporting some actions and identifying them as acceptable, as well as further providing possibilities for sanctioning non-compliance. The utilization of economic and financial instruments may thus be applied to further support actions. However, as they usually result in a cost to the state in terms of implementation e.g., of systems for fines or perhaps costs for grants, they are also unlikely to be applied in policy areas that are more newly emerged or on which less agreement currently exists [22].

Finally, soft instruments are characterized by being voluntary and non-coercive. Rather than imposing sanctions or requirements, they utilize persuasion to develop mutual agreements, and may consist of recommendations or codes of conduct, public-private partnerships, information campaigns or voluntary/contractual agreements [21]. These are the instruments by which emerging policy areas are likely to be governed, and, so far, this is the area in which climate issues have been situated. For instance, "Most countries have opted, at least up to now, for 'softer' forms of governing and have abstained from intervening directly in local climate politics" [22]. However, this is despite that "the economic literature suggests that, in some contexts, voluntary approaches can be effective if carefully designed but are not likely to be effective otherwise" [28]. Thus, while developing towards potentially larger institutionalization would be a natural development for issues that gain large salience, reaching this level of institutionalization takes time. Before issues gain sufficient consensus on means and methods to become institutionalized in legislation, less hierarchical, soft instruments remain subject to voluntary agreement among actors. Thus, for example, "'soft' policy instruments not backed by EU legislation encourage coordination, benchmarking and best practice without any threat of sanctions" [16]. In the climate area, these have included guidelines for local authorities and dissemination of information on good-practice cases [22]. Subsequently, among such actors, agreement and integration of these areas may be limited to a discursive or general policy agreement level, including a general agreement on the importance of measures (such as sustainable development or adaptation, the level so far highlighted as most common for agreement on adaptation according to [1]).

Institutionalization through soft measures, however, also result in actors who do not agree with these, or de-prioritize work in these areas due to other priorities or issues that are regarded as more pressing, only have normative and non-binding claims on them to integrate this area. This may be particularly noticeable at the local level. While local government may be able to self-organize in its capacity as a relatively strong actor within the planning framework and generally within the areas of municipal jurisdiction (e.g., education or health care), national level may pose multiple other requirements than climate in these areas—requirements that are mandatory. Thus, it may only be that after all other, stronger, requirements are taken into account that voluntary climate issues might be able to compete for priority (and then almost by necessity for a very limited policy space and resources). Issues that are not subject to binding regulation may thus well be de-prioritized as time and resources focus on mandatory issues (of which implementation of EU directives is one)—in particular, if funding is limited to project funding, which may be the case even in well-resourced (capital) cities [29].

The consideration with regard to climate may thus not be that municipalities or local government are not formally able to act on these issues but rather that the regulative environment and incentive structure may result in a situation where prioritizing voluntary aims at the level of or higher than mandatory ones may be difficult, given limited resources. However, a large number of potential considerations can formally be taken by local government. For instance, it has been suggested that urban climate governance may self-govern or govern by provision with regard to the activities it controls and that "strong leadership in the integration of processes has the ability to compensate for a lack of guidance or supporting legislation from higher decision-making levels" [30]. Local government may, for example, demand specific procurement within existing procurement-requirements already adopted by higher levels. It may also influence action through its ownership share in municipal companies, or act to "improve energy efficiency in government offices and other municipality-owned buildings" [22]. Another possibility is to take initiative for, or participate in, networks on e.g., urban climate change policy [18], leading to the utilization of legal and regulatory instruments at the local level, such as planning instruments [22], in relation, for instance, to adaptation. In a similar manner, interaction with the EU level could result in a sort of "download Europeanization" at the urban level by cities changing their "policies, practices, preferences or participants within local systems of governance, arising from the negotiation and implementation of EU programs", complemented by an "upload Europeanization" whereby "innovative urban practices" may result in the "incorporation of local initiatives in pan-European policies or programs" [20].

3.3 EXAMPLES OF IMPLEMENTATION IN THE NORDIC CONTEXT: CROWDING OF POLICY SPACE

The review of the use of instruments suggests that, as the policy arena matures and issues become more highly accepted, so may the use of instruments move from voluntary towards being institutionalized, and, if judged important, included within binding regulation (such as the climate laws that are now emerging in some European states). However, it would also appear that stronger regulation with regard to climate is found in terms of the more established areas of energy policy vis-à-vis mitigation, and flood management vis-à-vis adaptation.

In Denmark, combinations of actions mandated on the national and EU level are coordinated and partly prioritized at the local level. At the EU level, the EU Floods Directive (2007/60/EC) [31] has direct impact on the municipalities, especially those that have been identified as a risk area in Denmark under the Directive [32]. The Floods Directive requires extensive flood risk assessment analysis and action plans for all member states. A growing focus on climate change adaptation has been developed mainly in relation to large-scale flood event risk, considerations which resulted in a national requirement that all Danish municipalities must develop climate adaptation plans by the end of 2013 [33]. Subsequent changes in national law allow local utility companies the opportunity to invest in areas even if these are not directly related to waste water management—for instance, projects which prevent rainwater from reaching the sewage system, by using green areas, permeable surfaces and soakaways for retention and infiltration [34]. Another law opened up not only aesthetic or functional considerations but also climate-related issues to motivate restrictions in the compulsory local plans drawn up by municipalities [35]. The responsibility of practical climate adaptation at the local level has been designated to the individual municipality, where the role of the national government is to facilitate and control this process by providing information, advice and regulations. To this end, the Danish Ministry of the Environment is imposing a requirement that climate adaptation plans should, at a minimum, include risk assessment mapping and action plans developed in cooperation with the public and relevant stakeholders [36]. An advisory group of experts was created to visit municipalities and assist them in their climate adaptation planning. However, no specific funding exists beyond the opportunity to co-finance projects with utility companies. Large variations in how this legislation is implemented may exist across municipalities, as the level of institutionalization of climate change adaptation thereby centers on requiring development of climate change adaptation plans. Beyond that, it mainly provides possibility for, rather than mandates, consideration to climate (for example, allowing for climate-related issues to be included in compulsory local plans).

In Finland, adaptation has primarily been governed through soft policy instruments and has not resulted in any corresponding requirements for e.g., developing adaptation plans at the municipal level. Finland's National Adaptation Strategy, published as early as 2005 [37], proposed adaptation action on a sectoral basis but was not coupled to implementation requirements. The Strategy was considered to provide little guidance for local adaptation [38,39]. In addition, the recently published National Adaptation Plan for Climate Change 2022 [40] stresses the sectoral and individual responsibility for adaptation, while the state's role is restricted to safeguarding society's most central functions and the provision of expertise. The Finnish Climate Act, passed in 2015 (HE 82/2014) steers the government in the preparation of a national plan for adaptation and has details on the reporting. However, it does not have influence on the local adaptation, nor measures to steer it. So far, despite not being compelled in a legally binding sense to do so, some municipalities have developed adaptation strategies or have addressed mitigation and adaptation in a common strategy [41,42,43,44]. All of these strategies also remain soft instruments, since their implementation is not compulsory, nor is the non-implementation of the strategy penalized by sanctions or the withdrawal of rights. However, city departments can make binding decisions in their role as public authorities, and the consideration of new knowledge in local building codes or detailed plans can provide entry points for informed climate change adaptation [45]. Legislation also offers possibilities to shape climate change adaptation, especially when it comes to changes in the frequency and intensity of floods and precipitation patterns. However, few practical adaptation measures have been undertaken separately from flood protection measures based on the Land Use and Building Act requirement to promote a safe and healthy living environment (132/1999, Section 5).

Financial governing instruments that may promote climate change adaptation by private individuals and companies are not widely used. The insurance of homes against flood damage offers, in theory, such an instrument, but in practice to date, premiums do not depend on the house location [46].

Issues of implementation that are relevant for comparison can be highlighted with regard to a case of mitigation in Norway. We chose the case of mitigation in the private housing sector as it constitutes one of the least strongly regulated sectors of what is otherwise a relatively strongly regulated national mitigation regime. However, as it is situated within the broader context of a strongly regulated sector, voluntary mechanisms—for instance with regard to house owners renovations—could be seen as underpinned by a context of stronger demands.

Generally with regard to mitigation in the housing sector, Norwegian climate and energy policy broadly follows that of the EU. National strategic documents reflect EU concerns for reducing climate gas emissions from buildings, as well as the challenge posed by existing building mass and private dwellings which tend to be renovated outside the sphere of regulation [47,48]. While home renovation is steered with soft instruments, such as information campaigns, counselling services, economic support for components such as energy efficient heating systems and work such as insulation and campaigns aimed at producers of building components [49], binding regulations play an essential role in coordinating the efforts of policy agents at different levels. One example is the ban on oil heaters in houses from 2020 which was initiated by the EU and adopted by Norway. This requirement has given policy actors in Norway a fixed point to coordinate their efforts around. The ban is also one of the few regulations that directly impact renovation in private dwellings as it requires the replacement of oil heaters. It motivates house owners to show up at meetings, seek information about technologies, and support programs and solutions. The regulatory instrument also provides a context for the application of financial and soft instrument, bringing climate policy into private homes. Another example of the importance of legal and regulatory instruments is the building code in which Norway, to a great extent, follows EU requirements. As the code establishes a concrete, specific standard, it has a coordinating effect for the various agents of policy implementation including builders, producers of building components, do-it-yourself stores and energy consultants. As such, the building code runs like a line from the EU, through national policy, to municipalities, thus forcing actors implementing climate policy to coordinate with each other.

Thus, while the energy area is both highly regulated—since the 1970s as part of the first environmental policy wave—and exerts considerable impact on mitigation [50], this "loophole" area of private housing could be regarded as an exception in this broader context. It demonstrates the interacting complexities of a difficult-to-regulate subarea, difficult-to-reach individual actors, and soft regulation. At present, energy policy for private dwellings is implemented in practice by policy actors representing various levels of government, as well as by private institutions such as DIY stores, building companies and independent energy consultants. Many of them take part in an informal network on the national level [51], or in local level networks. At the municipality level, many local initiatives are underway aimed at energy efficiency in private dwellings, for example support to local energy counselling services and financial support for energy upgrading. Legal and regulatory instruments are coordinated between municipalities first and foremost in the processing of planning permission applications which follows the national building code. However, these instruments do not affect renovation projects in private dwellings directly as they are not regulated in this context. Soft and financial initiatives are thus not coordinated between municipalities in Norway but are increasingly coordinated through public enterprises operation at the national level, most notably Enova, an enterprise specifically set up to enhance the transition to greener energy at the national level, and the Housing Bank, providing affordable loans to house owners on the condition of ambitious energy standards [52]. As a result, a house owner engaged in a renovation project may experience that a number of policy agents have a say in the project, and they may provide different advice or express different requirements—and thus that an unclear policy situation with multiple potential choices limits action on mitigation in private housing. Even though policy agents tend to form networks in Norway, several actors have expressed their frustration over difficulties in coordinating their efforts, both between different levels of government but also between institutions at the same level.

3.4 DISCUSSION

This note discusses implementation of climate policy in the Nordic context. The three examples show how the policy arena is created through the multilevel governance arrangements, resulting in the proliferation of instruments for steering. The case of Norway illustrates well the problematic nature of nested implementation within the more highly regulated and more traditional environmental energy policy area. In this case of reduction of energy use in buildings, significant cooperation was seen as developing as a result of binding regulations, requiring coordination between actors. However, given the large number of actors in the sector and the complexity and subsequently modestly regulated nature of energy use in family level dwellings, this area was subject both to multiple and uncoordinated actors and soft regulation with limited implementation. The case can thus be seen to illustrate the fact that governance (between multiple actors and using multiple types of instruments, in particular soft instruments) requires much more coordination between actors than does traditional government steering in order to be successful [16]. The problems with coordination of climate policy for buildings are first and foremost concerned with financial and soft instruments, as agents do not coordinate efforts. This is because they are not forced to do so, and house owners turn away from the voluntary policies as its focus on details (in line with broader regulation, but applicable there as mandatory) contradicts their own more fluid renovation practices, and they are often not considered consistent.

In the two adaptation cases, in Denmark and Finland, flood management imposes by far the greater regulative requirements compared to adaptation to climate change. In comparison to flood management, adaptation policy has not gained such a level of implementation in the EU (where it is so far only developed as a non-binding strategy [53]), nor at the national level. In both the Danish and Finnish cases, thus, adaptation is largely managed through considerations in relation to flood risk under the existing EU requirements. This raises concerns over how to coordinate these requirements with the national adaptation plan requirements. The two countries diverge in that Denmark has national level requirements for municipal development of an adaptation plan, while a development of an adaptation plan in Finland is voluntary. However, the requirements for adaptation developed through legal means largely open up for municipalities to consider adaptation in Denmark, rather than require them to do so, and place no binding requirements on municipalities.

Hence, the regulation that is becoming most important is not in the emerging area of adaptation but focused on the case of flood risk. This "add on" of a new policy problem, a complex one, results in multiple questions when considering the choice of policy instruments. If this new area, adaptation in this case, consists of already regulated policy areas addressing parts of the new concern, how does one integrate new steering mechanisms in a way that does not reduce, contradict or in other ways "mis-match" with existing ones? These considerations suggest that the possibility of "mainstreaming" or integrating climate change into an existing regulative context can not be taken as given, but should rather be problematized with regard to what can be done at what level of regulation and use of instruments, within the existing regulative context and the competing priorities or requirements that may exist [54,55]. These findings are partly supported elsewhere, as it has been found that, despite significant possibilities for the municipal level to self-organize such as through providing leadership on climate issues or utilizing planning instruments available on the local level, measures have often retained a relatively soft, guiding character [1,22,30,56,57,58]. While successful examples can be seen for instance in the Netherlands, it could be argued that the high level of integration of water issues into legislative and planning systems [59] and their application in well-resourced larger cities, such as Rotterdam, as well as related existing spatial planning mechanisms and processes, together pose a context of well-developed instruments and practices without which the integration of adaptation in these cases cannot be understood [2].

While soft measures have often been highlighted as crucial to the local level, this comment thus calls to attention that there are inherent limits in such instruments, for instance in the absence of political will or given de-prioritization in relation to existing mandatory requirements [24]. Non-mandatory developments may be less likely to be well funded and will have to compete with all mandatory requirements for funding and implementation at any level [11,60]. Participation and bottom-up issue building will, while possibly locally effective, thus require high levels of agreement as well as coordination to attain uploading to European level [16].

3.5 CONCLUSIONS

Given that climate change has been emphasized as an area of broad societal risk, we thus highlight a need to develop approaches to integrate the emerging area of adaptation within a structure of mandatory regulation. These do not need to contradict the necessity to consider local variability [33]: first steps could, for example, include requiring explicit decisions on the role of adaptation in various council decisions where relevant. The role and application of mandatory adaptation planning now required in the Danish case may also be important to consider as a model for other countries. While it is difficult to pinpoint in what other specific ways stronger regulation could have been enacted in the Danish and Finnish cases, as multiple options exist, a comparison could be made with a UK case. In England, the integration of an indicator on adaptation within a national assessment system provided a reason for local government to prioritize adaptation, as adaptation could be selected as one of the indicators on which local government reported and was assessed for funding on the national level [60]. After the decommissioning of the indicator system and this national indicator with it, interviewees noted that it was becoming more difficult to prioritize adaptation [60,61] While the ways in which adaptation is more strongly regulated need thus not necessarily always be through legislation, increasing institutionalization will need to relate to existing systems providing for prioritarization and setting requirements. Relevant measures will depend on and should be assessed within the types of regulative systems that already exist, with an eye to placing climate change at a level that reflects the importance it is accorded in that decision-making system.

Rather than making the limited implementation of climate change issues so far surprise us, we should thus instead review the regulative context and opportunities [24]—not least given the large variety of areas with which climate change issues intersect—for integration or mainstreaming that places climate not only rhetorically but also in practice at par with other issue areas.

REFERENCES

1. IPCC. Climate Change 2014: Impacts, adaptation and vulnerability. Part A: Gloable and Sectoral Aspects. In Contribution of Working Group II to the Fifth Assessment Report of the Intergovernmental Panel on Climate Change; Field, C.B., Barros, V.R., Eds.; Cambridge University Press: New York, NY, USA, 2014; pp. 1–32.
2. Root, L.; van der Krabben, E.; Spit, T. Bridging the financial gap in climate adaptation: Dutch planning and land development through a new institutional lens. J. Environ. Plan. Manag. 2014, 58.
3. Urwin, K.; Jordan, A. Does public policy support or undermine climate change adaptation? Exploring policy interplay across different scales of governance. Glob. Environ. Change 2008, 18, 180–191.
4. Naess, L.O.; Bang, G.; Eriksen, S.; Vevatne, J. Institutional adaptation to climate change: Flood responses at the municipal level in Norway. Glob. Environ. Change 2005, 15, 125–138.
5. McEvoy, D.; Lindley, S.; Handley, J. Adaptation and mitigation in urban areas: Synergies and conflicts. Proc. ICE-Munic. Eng. 2006, 159, 185–191.
6. Casado-Asensio, J.; Steurer, R. Integrated strategies on sustainable development, climate change mitigation and adaptation in Western Europe: Communication rather than coordination. J. Public Policy 2014, 34, 437–473.
7. Lafferty, W.M.; Cohen, F. Conclusions and Perspectives. In Sustainable Communities in Europe; Lafferty, W.M., Ed.; Earthscan: London, UK, 2001; pp. 262–268.
8. Eckerberg, K. Sweden: Problems and prospects at the leading edge of LA21 implementation. In Sustainable Communities in Europe; Lafferty, W.M., Ed.; Earthscan: London, UK, 2001; pp. 15–40.
9. Rowe, J.; Fudge, C. Linking national sustainable development strategy and local implementation: A case study in Sweden. Local Environ. 2003, 8, 125–140.
10. Emilsson, S.; Hjelm, O. Towards sustainability management systems in three Swedish local authorities. Local Environ. 2009, 14, 721–732.
11. Keskitalo, E.C.H.; Liljenfeldt, J. Working with sustainability: Experiences of sustainability processes in Swedish municipalities. Nat. Resour. Forum 2012, 36, 16–27.
12. Black, J. Critical Reflections on Regulation. Aust. J. Legal Philos. 2002, 27, 1–36.
13. Braithwaite, J.; Coglianese, C.; Levi-Faur, D. Can regulation and governance make a difference? Regul. Gov. 2007, 1, 1–7.
14. Smith, D.K. Beyond the rule of law? Decentered regulation in online investing. Law Policy 2004, 26, 439–476.
15. Hooghe, L.; Marks, G. Types of Multi-Level Governance. Available online: http://eiop.or.at/eiop/pdf/2001-011.pdf (accessed on 12 October 2001).

16. Stephenson, P. Twenty years of multi-level governance: "Where does it come from? What is it? Where is it going?". J. Eur. Public Policy 2013, 20, 817–837.
17. Kaiser, R.; Prange, H. Managing diversity in a system of multi-level governance: The open method of co-ordination in innovation policy. J. Eur. Public Policy 2004, 11, 249–266.
18. Bulkeley, H.; Betsill, M.M. Revisiting the urban politics of climate change. Environ. Pol. 2013, 22, 136–154.
19. Gualini, E. Challenges to multi-level governance: Contradictions and conflicts in the Europeanization of Italian regional policy. J. Eur. Public Policy 2003, 10, 616–636.
20. Marshall, A. Europeanisation at the urban level: Local actors, institutions and the dynamics of multi-level interaction. J. Eur. Public Policy 2004, 23, 668–686.
21. Borrás, S.; Edquist, C. The choice of innovation policy instruments. Technol. Forecast. Soc. Change 2013, 80, 1513–1522.
22. Kern, K.; Alber, G. Governing climate change in cities: Modes of urban climate governance in multi-level systems. In Proceedings of the Competitive Cities and Climate Change OECD Conference, Milan, Italy, 9–10 October 2008; pp. 171–196.
23. Jordan, A.; Wurzel, R.K.W.; Zito, A. The rise of "new"policy instruments in comparative perspective: Has governance eclipsed government? Polit. Stud. 2005, 53, 477–496.
24. Lascoumes, P.; Le Gales, P. Introduction: Understanding public policy through its instruments—From the nature of instruments to the sociology of public policy instrumentation. Governance 2007, 20, 1–21.
25. Appelstrand, M. Miljömålet i Skogsbruket—Styrning och Frivillighet; Lunds Studies in Sociology of Law, University of Lund: Lund, Sweden, 2007.
26. Nilsson, S. Experiences of policy reforms of the forest sector in transition and other countries. For. Policy Econ. 2005, 7, 831–847.
27. Kassim, H.; Le Galès, P. Exploring governance in a multi-level polity: A policy instruments approach. West. Eur. Politics 2010, 33, 1–21.
28. Segerson, K. Voluntary approaches to environmental protection and resource management. Annu. Rev. Resour. Econ. 2013, 5, 161–180.
29. European Environment Agency (EEA). Urban Adaptation to Climate Change in Europe: Challenges and Opportunities for Cities Together with Supportive National and European Policies; Report No. 2/2012; EEA: Copenhagen, Denmark, 2012.
30. Wamsler, C.; Luederitz, C.; Brink, E. Local levers for change: Mainstreaming ecosystem-based adaptation into municipal planning to foster sustainability transitions. Glob. Environ. Change 2014, 29, 189–201.
31. European Commission. Directive 2007/60/EC of the European Union and the Council on the Assessment and Management of Flood Risks. Available online: http://eur-lex.europa.eu/legal-content/EN/TXT/?uri=CELEX:32007L0060 (accessed on 23 October 2007).
32. Endelig Udpegning af Risikoområder for Oversvommelse fra Vandlob, Soer, Havet og Fjorde. Available online: http://naturstyrelsen.dk/media/nst/Attachments/Udpegningafrisikoomrder_NSTKDI.pdf (accessed on 18 December 2015).
33. Aftale om Kommunernes Okonomi for 2013. Available online: http://www.kl.dk/ImageVaultFiles/id_55201/cf_202/Aftale_om_kommunernes_%C3%B8konomi_for_2013.PDF (accessed on 10 June 2015).

34. Lov om Ændring af lov om Betalingsregler for Spildevandsforsyningsselskaber m.v. og lov om Vandlob. Available online: https://www.retsinformation.dk/forms/r0710.aspx?id=145194 (accessed on 29 January 2013).
35. Erhvervs-og, V. Chapter 4: Kommuneplanlægning. In Bekendtgorelse af Lov om Planlægning; Available online: http://www.viauc.dk/sygeplejerske/holstebro/Documents/Intranet/retningslinjer%20for%20skriftlige%20opgaver%20rev%20januar%202012.pdf (accessed on 23 November 2015).
36. Klimatilpasningsplaner og klimalokalplaner—Vejledning. Available online: http://docplayer.dk/3776-Klimatilpasningsplaner-og-klimalokalplaner-vejledning.html (accessed on 18 December 2015).
37. Marttila, V. Finland's National Strategy for Adaptation to Climate Change; Ministry of Agriculture and Forestry of Finland: Helsinki, Finland, 2005.
38. Evaluation of the Implementation of Finland's National Strategy for Adaptation to Climate Change 2009. Available online: http://mmm.fi/documents/1410837/1721034/Adaptation_Strategy_evaluation.pdf/043c0964-58c5-4fce-8924-cc47748cf766 (accessed on 18 December 2015).
39. Juhola, S.; Haanpää, S.; Peltonen, L. Regional challenges of climate change adaptation in Finland: Examining the ability to adapt in the absence of national level steering. Local Environ. 2012, 17, 629–639.
40. Kansallinen Ilmastonmuutokseen Sopeutumissuunnitelma 2022. Available online: http://mmm.fi/documents/1410837/1516663/2014_5_lmastonmuutos.pdf/1716aa76-8005-4626-bae0-b91f3b0c6396 (accessed on 20 November 2014).
41. Helsingin Seudun Ympäristöpalvelut-Kuntayhtymä (HSY) Pääkaupunkiseudun Ilmastonmuutokseen Sopeutumisen Strategia. Available online: https://www.hsy.fi/sites/Esitteet/EsitteetKatalogi/Julkaisusarja/10_2012_paakaupunkiseudun_ilmastonmuutokseen_sopeutumisen_strategia.pdf (accessed on 20 April 2015).
42. Helsingin Kaupungin Ympäristöpolitiikka, Kaupunginvaltuuston Hyväksymä. Available online: http://www.hel.fi/static/ymk/esitteet/ymparistopolitiikka.pdf (accessed on 26 September 2012).
43. Keski-Pohjanmaan Ilmastostrategia 2012–2020. Available online: https://www.kokkola.fi/palvelut/kaavat_ja_kiinteistot/kaavoitus/yleiskaavoitus/kokkolan_keskustaajaman_yk/perusselvitykset_keskustaajama/fi_FI/selv_ymparistohaitat/_files/92461515146004114/default/Keski_Pohjanmaan_ilmastostrategia_2012_2020.pdf (accessed on 3 November 2011).
44. Oulun Seudun Ilmastostrategia. Available online: http://www.ouka.fi/documents/64417/6b35350e-08ef-411a-b904-abfa45d89544 (accessed on 19 May 2012).
45. Klein, J. Embeddedness of climate change adaptation: Established procedures and contending discourses for flood protection in Espoo, Finland. Local Environ. 2014, 8, 1–18.
46. Tulvat Hiipivät; Kotivakuutuksiin. Available online: http://www.hs.fi/kotimaa/Tulvat+hiipiv%C3%A4t+kotivakuutuksiin/a1402465110828 (accessed on 12 June 2014).
47. Meld. St. 28 (2011–2012): Gode Bygg for Eit Betre Samfunn. Available online: https://www.regjeringen.no/no/dokumenter/meld-st-28-20112012/id685179/?ch=1&q (accessed on 18 December 2015).

48. Lavenergiutvalget: Energieffektivisering. Available online: http://www.regjeringen.no/upload/OED/Rapporter/OED_Energieffektivisering_Lavopp.pdf (accessed on 18 December 2015).
49. Solli, J.; Fyhn, H. Socialising energy policy for home renovation. Energy Policy Soc. Sci. 2016. in press.
50. Liefferink, D.; Andersen, M.S. Strategies of the "green" Member States in EU environmental policy-making. J. Eur. Public Policy 1998, 5, 254–270.
51. Simon, G.; Shove, E. The Sociology of Energy, Buildings and the Environment: Constructing Knowledge, Designing Practice; Routledge: London, UK, 2004.
52. Potensial-og Barrierestudie: Energieffektivisering i Norsk Bygg. Available online: http://www.enova.no/upload_images/A319D657AFC34028B18D5C74BA4F0094.pdf (accessed on 18 December 2015).
53. The EU Strategy on Adaptation to Climate Change. Available online: http://ec.europa.eu/clima/publications/docs/eu_strategy_en.pdf (accessed on 30 July 2013).
54. Ayers, J.M.; Huq, S.; Faisal, A.M.; Hussain, S.T. Mainstreaming climate change adaptation into development: A case study of Bangladesh. WIREs Clim. Change 2014, 5, 37–51.
55. Uittenbroek, C.J.; Janssen-Jansen, L.B.; Runhaar, H.A.C. Mainstreaming climate adaptation into urban planning: Overcoming barriers, seizing opportunities and evaluating the results in two Dutch case studies. Reg. Environ. Change 2013, 13, 399–411.
56. Wejs, A. Integrating climate change into governance at the municipal scale: An institutional perspective on practices in Denmark. Environ. Plann. C Gov. Policy 2014, 32, 1017–1035.
57. Lonkila, K. Aspects of Strategic Climate Work in Nordic Municipalities: NordLead Project Final Report; Nordic Council of Ministers: TemaNord, Helsinki, Finland, 2012.
58. Mossberg Sonnek, K.; Johansson, B.; Lindgren, J. Risk and vulnerability analysis: A feasible process for local climate adaptation in Sweden? Local Environ. 2013, 18, 781–800.
59. Scholten, P. Dutch approaches to flood risks: Developments in integrative water management and the synchronization of public and private agendas for climate adaptation in the Netherlands. In Climate Change and Flood Risk Management: Adaptation and Extreme Events at Local Level; Keskitalo, E.C.H., Ed.; Edward Elgar Publishing: Gloucestershire, UK, 2013; pp. 258–289.
60. Dupuis, J.; Keskitalo, E.C.H.; University of Lausanne, Lausanne, Switzerland. unpublished work. 2014.
61. Keskitalo, E.C.H. Climate change adaptation in the United Kingdom: England and South-East England. In The Development of Adaptation Policy and Practice in Europe: Multi-Level Governance of Climate Change; Keskitalo, E.C.H., Ed.; Springer: Dordrecht, The Netherlands, 2010; pp. 97–147.

CHAPTER 4

Accounting for Greenhouse Gas Emissions of Materials at the Urban Scale-Relating Existing Process Life Cycle Assessment Studies to Urban Material and Waste Composition

MEIDAD KISSINGER, CORNELIA SUSSMANN, JENNIE MOORE, AND WILLIAM E. REES

4.1 INTRODUCTION

It has been estimated that about 78% of global carbon emissions can be directly and indirectly related to cities [1,2]. To avoid the most catastrophic consequences of global climate change, greenhouse gas (GHG) emissions associated with urban centres must be dramatically reduced [3-5]. Toward this goal, many cities are engaged in efforts to calculate and reduce their green house gas emissions. Most are accounting for GHGs produced within urban boundaries, often referred to as "scope one" emissions. A growing awareness among researchers suggests that in order to achieve globally relevant reductions in atmospheric carbon levels, municipal governments and urban residents should also take responsibility for urban "lifestyle" or consumption emissions: emissions mostly related to the GHGs embodied in the life cycle of material goods (as well as food) consumed in the city (scope 3 emissions) [6-10]. Information about the GHG emissions associated with the manufacturing of specific materials can be used to generate public awareness about implications of material consumption choices and habits. It can also be used to develop municipal policies and programs targeting high-emissions materials for reduction. However, data constraints and GHG accounting complexity present challenges. We suggest that cities can use the solid waste life cycle assessment (LCA) based approach to account for their material consumption. We used the approach to identify a set of materials commonly con- sumed in cities, and then reviewed published LCA data to develop a range of GHG emissions values for each material. Our review of the studies and our dataset of emissions values are presented.

Cities that measure their GHG emissions follow international protocols such as the International Local Government GHG Emissions Analysis Protocol [11]. These protocols account for emissions perceived to be directly within the control of the local government. "Scope one" includes emissions from facilities that are owned by the local government or emissions produced by citizens' activities within city limits, for example, from motor vehicle transportation. Emissions associated with electrical energy used to operate buildings and emissions from solid waste management are also counted, even though these emissions are sometimes generated outside the city, e.g. at a remote power station.

Several studies have followed similar principals in generating GHG emissions inventories for urban settlements [12-15]. Bi et al. [14] produced a bottom up GHG emissions inventory for Nanjing, China. They included emissions from industrial, transport, commercial and household energy consumption; emissions from industrial processes located within the city, and emissions from waste treatment. Kennedy et al. [13] generated GHG emissions inventories for ten cities on four continents. Their method includes seven components: electricity, heating and industrial fuels, industrial processes, ground transportation, aviation transportation, marine transportation, and waste.

Few researchers have conducted studies that include urban consumption related or, scope 3, emissions. One challenge has been data limitations [15]. Hillman and Ramaswami [16] calculated GHG emissions for eight US cities including embodied emissions in food, transport fuels, shelter and cross-border freight demands. Yang and Suh [17] accounted for the GHG emissions related to products consumed by Chinese urban and rural households; and Druckman and Jackson [18] calculated the GHG emissions required to satisfy average UK household demand for goods and services between 1990 and 2004. To date, no standard method for assessing GHG emissions from urban material consumption has been determined.

One GHG accounting approach increasingly being used at the sub-national/urban scale is "environmental input-output analysis" (EIOA) [19-21]. It uses local expenditure data ($) for some consumption items like food and materials, and relates them to carbon emissions in an extension of conventional monetary input-output analysis. However, that approach usually does not provide data at the scale of specific material types such as newsprint and cardboard, or even at the scale of product groups like paper or plastic. Rather, EIOA operates at the industry scale (e.g., emissions per $ value of the national paper or plastic industry). Further, the approach requires cities to have detailed residents' expenditure data to generate input-output tables, a requirement that many cities cannot easily meet.

It follows that if cities are to take on measurement, monitoring and development of policies to reduce material consumption-related GHG emissions, they require local data and a method that is not too onerous [22]. The "solid waste LCA based approach" [23-26] we suggest here overcomes data limitations by using data many cities already collect, solid waste volume and composition data, to identify patterns of material consumption. It then uses data from a wide range of life cycle assessment studies to determine the GHG emissions associated with production of a material or product. For this paper, we used the approach to identify fourteen materials commonly consumed in cities in high income countries, and conducted a thorough review of published, process life cycle studies to determine GHG emissions values for each material. The range of GHG emissions values we present for each material reflects the variability of life cycle characteristics associated with production method and location.

4.2 METHODS

While cities do not commonly monitor or document their residents' material consumption, they do manage and monitor solid waste. The "component solid waste LCA based approach" uses urban waste stream data to identify the major types of materials consumed in urban areas. This approach to estimating urban material consumption was developed by Simmons et al. [24]; Chambers et al. [26], and Barrett et al. [23] as part of their studies on urban sustainability using ecological footprint analysis. It has been used since by some footprint studies at the urban scale [21,22]. The logic behind the approach is that most materials consumed end up in the waste stream, some in a matter of minutes after consumption, others after years. Although waste stream data will not represent the exact quantities of all materials consumed in a city over a given period of time, it is reasonable to assume that the proportions of materials (by weight) found in the waste stream reflect the proportions consumed. In this way, a set of regularly consumed materials can be identified by type and weight. In absence of other urban material consumption data, the waste stream serves as a useful proxy.

We reviewed waste stream documentation and reporting protocols for ten cities in relatively high income countries: Canada, United Kingdom, the United States, Australia and Israel. The purpose of the review was to identify a general trend in the way solid waste is documented, and to generate a list of commonly reported waste items. Cities that monitor and document comercial and household waste composition collect data on the following major categories: metal; glass; plastics; paper; organics; textiles; rubber; and hazardous wastes. Many use more detailed categories. For example, paper is broken down into paper, newsprint, and cardboard. Plastics are identified by type (polyethylene terephthalate [PET]; high density polyethylene [HDPE]; low density polyethylene [LDPE] and polyvinyl chloride [PVC]) and by use such as plastic (film) bags and plastic bottles (e.g., Sydney, AU, 2009; Vancouver, CA, 2010; Seattle, USA, 2010; Edinburgh UK, 2010). One consumer item that appears in the solid waste stream at high volume and is commonly reported as a separate item is diapers (nappies). In both Sydney, Australia and various cities in Israel [27, 28], diapers made up approximately 5%, by weight, of the residential waste stream.

For our materials dataset, we selected the fourteen materials most commonly reported in the urban waste stream data we reviewed: paper 1) newsprint, 2) print paper, 3) cardboard, plastics, 4) PET, 5) HDPE, 6) LDPE, 7) PS, 8) PVC, 9) steel, 10) aluminum, 11) glass, textiles, 12) cotton fabric, 13) polyester fabric, and 14) diapers.

The component solid waste LCA based approach draws GHG emissions data from process life cycle assessments. We conducted an extensive review of LCA studies and reports for each of the fourteen materials. The review generated a total of 120 values from 69 sources. For the complete list of studies and their emissions data see Appendix I. From each study, for each material, we extracted the GHG emissions (CO_2e) data.

Our literature review included LCA studies in academic literature, and in commercial and industrial publications. The studies include data from European, North American, Asian, and Australian sources among others to reflect world-wide production systems and conditions. Each LCA study sets its own boundaries and scale. In order to present comparable emissions values we made an effort to include studies that used similar parameters, assumptions, and scales. Overall we made an effort to cover cradle to gate data. This means data associated with the manufacturing process from materials extraction to finished product that leaves the factory gate. This approach avoids double-counting the energy and materials associated with the end of life cycle in which products are managed as wastes and for which local governments also maintain records through their regular waste management functions. In the case of plastics most of our values are for plastic polymers owing to lack of available LCA data on finished products. Our review of studies published in Chinese yielded few results. For most products we have only one data source from China.

Because the component solid waste LCA based approach relies on emissions data from LCA studies, it is limited by the availability and accuracy of those studies. LCA is well established in academic and private industrial realms, but comparability and credibility of LCA studies requires improvement [29]. Several bodies are working to improve standardization; for example, the European Commission project, European Platform on Life-cycle Assessment, resulted in a handbook of recommendations for life cycle impact assessment in Europe [30]. Continued standardization of

LCA protocols will benefit cities that choose to account for consumption related emissions using LCA based approaches.

4.3 RESULTS

Tables 1 and 2 summarize our GHG emissions review. Table 1 shows the minimum, maximum, mean and the standard deviation of emissions for materials in ascending order by type: glass; paper products; plastics; steel; diapers; aluminum; textiles. N represents the number of studies from which data was collected for each material. The table displays the relative GHG emissions among materials by unit of material (per tonne).

However, it is the total amount consumed that determines the actual impact of a material on the urban GHG emissions. Textiles and aluminum generate the highest GHG emissions per unit of material, but they represent a relatively smaller part of the overall weight of the urban waste stream (or consumption) in cities we reviewed.

Paper products have relatively lower GHG emissions per tonne, but comprise a significant proportion of many urban commercial and residential waste streams. For urban planners and policy makers, both the GHG emissions associated with a material's per unit production, and the total, on-going quantities consumed are necessary data. With these data, policies and programs can be directed toward reducing consumption of materials with high aggregate impact.

In our review of LCA studies we found variations in material emissions values for studies conducted in different parts of the world (Table 2). While these variations can be explained by variations in LCA methods and data availability, they also likely reflect characteristics of local production methods and energy sources (e.g., coal based electricity vs hydroelectric sources). As more LCA data from countries like China become available, these variations may be more prominently expressed.

A database of GHG emissions for use by city governments around the world could make accounting for urban consumption emissions a feasible endeavor. To account for variability in GHG emissions asso ciated with production location, such a database could provide an average value for each material that reflects the range of countries in which the

Table 1. Range of life cycle GHG emissions associated with materials, "cradle to gate".

	N	Min	Max	Mean	Standard Deviation
Sub Category		Kg CO_2e/t	Kg CO_2e/t	Kg CO_2e/t	Kg CO_2e/t
Glass	8	600	1800	990	370
Cardboard	9	560	1620	890	330
Newsprint	8	780	1670	1120	350
Printing Paper	15	420	3110	1290	770
HDPE	6	580	1950	1015	670
PVC	6	1400	2510	1920	370
PET	8	1070	2890	2240	600
LDPE	6	1870	2760	2360	380
PS	8	1180	4660	2970	1120
Steel	20	1600	4020	2530	730
Diapers	3	2600	4390	3580	900
Aluminum	9	7900	18,180	10,840	3170
Cotton Fabric	9	12,760	30,000	21,500	6770
Polyester Fabric	5	15,120	32,500	26,200	9600

goods are produced. The average could even reflect each nation's proportion of the global production market for individual materials. The database would also report minimum and maximum emission values. Cities could choose minimum, average or maximum emissions data for on-going monitoring.

4.4 CONCLUSIONS

Among researchers, efforts are being made to overcome data challenges, and account for scope 3 emissions, i.e., those associated with the embodied energy of material goods that are consumed within cities. The use of waste

Table 2. CO_2e and CO_2 emissions of materials by production location.

	Europe		America		Asia		Australia	
	Kg CO_2e/t	Kg CO_2e/t	Kg CO_2e/t	Kg CO_2e/t	Kg CO_2e/t	Kg CO_2e/t	Kg CO_2e/t	Kg CO_2e/t
Glass	600 - 1047	550 - 940	n/a	585 - 1250	n/a	1820	n/a	765
Printing Paper	830 - 1560	420 - 1460	520 - 1600	1410	n/a	2480	2200 - 3000	n/a
Newsprint	720 - 1230	784 - 1230	n/a	n/a	n/a	n/a	n/a	n/a
Cardboard	557 - 1080	615 - 990	580 - 3140	n/a	330 - 1600	n/a	1600	n/a
PET	2780 - 3480	1890 - 3700	1810 - 2660	1070 - 2330	n/a	1340	n/a	n/a
PVC	2280 - 3860	2150 - 3600	2740 - 3480	2390 - 3060	1770	1410	n/a	n/a
Polystyrene	4080 - 4860	3250 - 3990	n/a	n/a	n/a	3390	n/a	n/a
HDPE	2315 - 3430	510 - 2980	1080 - 3270	1010 - 2770	n/a	2030	1970	n/a
LDPE	2700 - 3590	2250 - 3100	3330	2820	n/a	1860	2760	n/a
Aluminum	8670 - 15,400	680 - 12,400	7100 - 10,700	7940 - 12,000	21,500 - 22,500	18,180	16,300 - 22,400	n/a
Steel	1800 - 2430	1700 - 3570	n/a	1560 - 2670	n/a	1720 - 3750	2300 - 6800	n/a
Cotton Textile	n/a	6500	n/a	7700 - 16,000	n/a	12,700-16,240	25,000	15,700
Polyester	n/a	5000	n/a	n/a	n/a	15,120	31,000	20,000 - 32,500

as a proxy for material consumption overcomes a major limitation of data availability for urban planners and policy-makers. Still, it is important to acknowledge that the solid waste LCA based approach probably does not capture the entire volume of materials consumed, and that the approach is highly dependent on the quality and specificity of solid waste data collection and documentation. Further, determination of GHG emissions values for materials depends on the quantity and quality of accessible, published LCA studies.

Our review of process life cycle assessment studies revealed that only a limited number use detailed, primary data. The literature is also lacking in studies related to production in China, a major manufacturer. Despite these gaps we were able to generate a range of GHG emissions values for each of fourteen materials commonly consumed in cities. Among these materials we found that textiles and aluminum are associated with relatively high GHG emissions per tonne of production, compared to other materials such as paper and plastics. However, paper and plastics are consumed (found in the waste stream) in higher quantities, by weight, than aluminum and textiles so they could have equal or greater impact on overall consumption-related emissions. Cities aiming to account for consumption-related emissions and to develop programs to reduce high impact material consumption need material-specific information on both GHG emissions per unit of production, and overall quantities of consumed.

We found an increasing number of material LCAs are being conducted or commissioned by commercial and industrial associations such as the World Aluminum Association, the European Container Glass Federation or the European plastic producers association. Individual companies are also publishing information on the GHG emissions of their products. Perhaps more industry based studies will become available as carbon taxes and cap and trade systems are expanded. Consumer pressure for more ecologically benign products may also encourage more reporting.

We see the material LCA approach as a valuable, accessible approach for cities working to assess, monitor and develop policy to reduce their consumption based contributions to global GHG emissions.

REFERENCES

1. G. L. R. Brown, "Eco-Economy: Building an Economy for the Earth," Norton, New York, 2001.
2. N. Grimm, S. Faeth, N. Golubiewski, C. Redman, J. Wu, X. Bai and J. Briggs, "Global Change and the Ecology of Cities," Science, Vol. 319, No. 5864, 2008, pp. 756-760. doi:10.1126/science.1150195
3. W. E. Rees, "Cities as Dissipative Structures: Global Change and the Vulnerability of Urban Civilization," In: M. P. Weinstein and R. E. Turner, Eds., Sustainability Sc i- ence: The Emerging Paradigm and the Urban Environ- ment, Springer, New York, 2012, pp. 247-273.
4. World Bank, "Cities and Climate Change: An Urgent Agenda," Urban Development Series Knowledge Papers, Vol. 10, International Bank for Reconstruction and De- velopment, World Bank, Washington DC, 2010.
5. P. Newman, T. Beatley and H. Boyer, "Resilient Cities: Responding to Peak Oil and Climate Change," Island P re s s, Washington DC, 2009.
6. J. Moore, "Measuring Cli mate Action in Vancouver: Co m- paring a City's Greenhouse Gas Emissions Inventory Pro- tocol to the Inventory of Consumption," In: B. Richard- son, Ed., Local Climate Change Law: Environmental Re- gulation in Cities and Other Localities, Edward Elgar Publishing Ltd., Cheltenham, 2012.
7. D. Dodman, "Blaming Cities for Climate Change? An Analysis of Urban Greenhouse Gas Emissions Invento- ries," Environment and Urbanization, Vol . 21, No. 1, 2009, pp. 185-201. doi:10.1177/0956247809103016
8. S. Bastioni, F. Pulsellini and E. Tiezzi, "The Problem of Assigning Responsibility for Greenhouse Gas Emissions," Ecological Economics, Vol. 49, No. 4, 2004, pp. 253-257. doi:10.1016/j.ecolecon.2004.01.018
9. J. Feng, "Allocating the Responsibility of CO2 Over-Emis- sions from the Perspectives of Benefit Principle and Eco- logical Deficit," Ecological Economics, Vol. 46, No. 1, 2003, pp. 121-141. doi:10.1016/S0921-8009(03)00104-6
10. J. Munksgard and K. A. Pedersen, "CO2 Accounts for Open Economies: Producer or Consumer Responsibility?" Energy Polic y, Vol. 29, No. 4, 2001, pp. 327-334. doi:10.1016/S0301-4215(00)00120-8
11. ICLEI, "International Local Government Greenhouse Gas Emissions Analysis Protocol—Version 1.0," 2009.
12. R. Barthelmie, S. Morris and P. Schechter, "Carbon Neu- tral Biggar: Calculating the Community Carbon Footprint and Renewable Energy Options for Footprint Reduction," Sustainability Science, Vol. 3, No. 2, 2008, pp. 267-282. doi:10.1007/s11625-008-0059-8
13. C. Kennedy, J. Steinberger, B. Gasson, Y. Hansen, T, Hillman, M. Havranek, D. Pataki, D. Phdungsilp, A. Ra- maswami and G. Mendez, "Greenhouse Gas Emissions from Global Cities," Environmental Science and Tech-nology, Vol. 43, No. 19, 2009, pp. 7297-7302. doi:10.1021/es900213p
14. J. Bi, R. Zhang, H. Wang, M. Liu and Y. Wu, "The Benchmarks of Carbon Emissions and Policy Implica- tions for China's Cities: Case of Nanjing," Energy Policy, Vol. 39, No. 9, 2011, pp. 4785-4794. doi:10.1016/j.enpol.2011.06.045

15. C. Kennedy, J. Cuddihy and J. Engel-Yan, "The Chang- ing Metabolism of Cities," Journal of Industrial Ecology, Vol. 11, No. 2, 2007, pp. 43-59. doi:10.1162/jie.2007.1107
16. T. Hillman and A. Ramaswami, "Greenhouse Gas Emis- sion Footprints and Energy Use Benchmarks for Eight US Cities," Environmental Science and Technology, Vol. 44, No. 6, 2010, pp.1902-1910. doi:10.1021/es9024194
17. Y. Yang and S. Suh, "Environmental Impacts of Prod- ucts in China," Environmental Science and Technology, Vol. 45, No. 14, 2011, pp. 4102-4109. doi:10.1021/es103206g
18. A. Druckman and T. Jackson, "The Carbon Footprint of UK Households 1990-2004: A Socio-economically Dis- aggregated, Quasi-multi-regional Input-output Model," Eco- logical Economics, Vol. 68, No. 6, 2009, pp. 2066-2077. doi:10.1016/j.ecolecon.2009.01.013
19. M. Lenzen and G. Peters, "How City Dwellers Affect Their Resource Hinterland—A Spatial Impact Study of Australian Household," Journal of Industrial Ecology, Vol. 14, No. 1, 2009, pp. 73-90. doi:10.1111/j.1530-9290.2009.00190.x
20. J. Munksgaard, M. Wier, M. Lenzen and C. Dey, "Using Input-Output Analysis to Measure the Environme ntal Pr es - sure of Consumption at Different Spatial Levels," Journal of Industrial Ecology, Vol. 9, No. 1-2, 2005, pp. 169-185. doi:10.1162/1088198054084699
21. J. Moore, M. Kissinger and W. E. Rees, "An Urban Metabolism Assessment of Metro Vancouver," Journal of Environmental Management, in Press.
22. M. Kissinger and A. Haim, "Urban Hinterlands: The Case of an Israeli Town Ecological Footprint," Environment Development and Sustainability, Vol. 10, No. 4, 2008, pp. 391-405. doi:10.1007/s10668-006-9071-2
23. J. Barrett, H. Vallack, A. Jones and G. Haq, "A Material Flow Analysis and Ecological Footprint of York," Stock- holm Environmental Institute, Stockholm, 2001.
24. C. Simmons, K. Lewis and J. Barrett, "Two Feet—Two Approaches: A Component-Based Model of Ecological Footprinting," Ecological Economics, Vo l . 32, N o . 3 , 2000, pp. 375-380.
25. C. Simmons and N. Chambers, "Footprinting UK House- holds: How Big Is Your Ecological Garden?" Local En- vironment, Vol. 3, No. 3, 1998, pp. 355-362. doi:10.1080/13549839808725571
26. N. Chambers, C. Simmons and M. Wackernagel, "Shar- ing Natures Interest," Earthscan, London, 2000.
27. APC, "Kerbside Domestic Waste and Recycling Audit," APC Environmental Management, Sydney, 2009.
28. IMEP, "Israel Ministry of Environmental Protection, 2005 Waste Composition Study," 2006.
29. B. Vigon, "Euro pean Platform on Life cy cle Assessment ," Society of Environmental Toxicology and Chemistry, Vol. 7, No. 2, 2006, pp. 39-40.
30. European Commission—Joint Research Centre—Institute for Environment and Sustainability, "International Refer- ence Life Cycle Data System (ILDC) Handbook—Rec- ommendations for Life Cycle Impact Assessment in the European Context," Publications Office of the European Union, Luxemburg, 2011.

There is a supplemental file that is not available in this version of the article. To view this additional information, please use the citation on the first page of this chapter.

CHAPTER 5

Does Size Matter? Scaling of CO_2 Emissions and U.S. Urban Areas

MICHAIL FRAGKIAS, JOSÉ LOBO, DEBORAH STRUMSKY, AND KAREN C. SETO

5.1 INTRODUCTION

Urbanization is a hallmark of the 21st century, characterized by massive demographic shifts and large-scale rapid expansion of urban areas and the built environment [1]. Recent estimates show that 60–80% of final energy use globally is consumed by urban areas [2] and more than 70% of global greenhouse gas emissions are produced within urban areas [3]. The majority of future population growth for the remainder of this century will occur in urban areas [4]. The increase in global energy consumption, due to a rise in population and wealth will have significant effects on greenhouse gas emissions, human wellbeing, and sustainability [5]–[6].

It is a stylized fact that cities offer benefits from economies of scale. The concentration of people, large scale infrastructure and economic activity enable innovation and efficiencies [7]. Per capita urban energy consumption in industrialized countries is often lower than national averages [8]. Several studies show that compact and mixed urban land use coupled with co-located high residential and employment densities can reduce energy consumption and emissions through reducing vehicle miles traveled [9]–[10]. In this paper, we examine the relationship between population size of cities and carbon dioxide (CO_2) emissions using data from the U.S. urban system.

One of the most salient characteristics of an urban area is it population size as it is both determinant and consequent of the socio-economic activity occurring within cities [11]. Urban population size has attracted significant attention across different disciplines as an indicator of the city and an explanandum of urban phenomena. A large body of literature in economics shows that larger urban agglomerations are more productive [7], [12] and more innovative [13]–[16]. The positive and strong relationship between urban size and productivity appears to be central characteristic of modern urban economies [17]. The importance of population size as a major factor in determining the intensity of socio-economic activity in urban areas has recently been emphasized by research that applies scaling analysis to a diverse spectrum of urban indicators [11], [18]–[19]. Scaling analysis, which has been a powerful tool across many scientific domains, represents how measurable aggregate characteristics respond to a change in the size of the system. Its analytical strength stems from the observation that this response is often a simple, regular, and systematic function over a wide range of sizes, indicating that there are underlying generic constraints at work on the system as it grows.

The population size of a city, as well as its spatial organization and structure can influence energy consumption. Energy is needed to both maintain existing infrastructure and to fuel economic activity while economic activity in turn affects energy demand [20], [21]. Calculations using a production-based accounting estimate that urban areas contribute approximately 30–40% of total anthropogenic greenhouse emissions - while, in contrast, a consumption-based accounting puts urban contributions at 60% of total, with a few wealthy cities contributing a majority of the emissions [8], [19], [22]. Data from world cities suggest that climate, technology, density and wealth are important determinants of energy use and CO_2 emissions [23]. Past research has also shown that cities with larger populations present advantages over smaller cities in terms of their energy efficiency and CO_2 emissions [24].

In this paper we examine the relationship between urban population size and urban CO_2 emissions and ask the question: Are larger cities more emissions efficient than smaller ones? Furthermore, what is the relative importance of population size compared to other determinants of emissions discussed above? Given that urban populations will increase by 2–3 billion by the end of the 21st century, understanding how urban size affects emissions can offer insight into how city size can be part of a larger regional or national strategy for reducing emissions. If larger cities are emissions efficient, national urban policy could encourage the development of large cities ceteris paribus—social, economic, and governance issues aside. Of course, urban and development policies would be constrained by other goals that cities—especially those in developing countries—are trying to achieve, including pollution abatement, poverty reduction, and industrialization, among others. Nonetheless, without fundamental scientific understanding of the relationship between urban population size and urban emissions, it is difficult for cities and national governments to prioritize sustainability and urbanization policies.

5.1.1 THE IMPORTANCE OF SCALE FOR URBAN CO_2 EMISSIONS

Scaling characterizes how a given systemic quantity of interest, Y, depends on the size of a system. A common feature of scaling is scale invariance, formalized as:

$$Y(N) = Y_0 N^{\beta} \quad (1)$$

where Y_0 is a normalization constant and β is the scaling exponent, which can also be interpreted as an elasticity as usually defined in economics [25]. The significance of this "power law" relation becomes evident when we consider an arbitrary scale change by a factor λ from N to λN. This induces a change in Y from Y(N) to Y(λN) that can be expressed as

$$Y(\lambda N) = Z(\lambda, N)Y(N) \tag{2}$$

This equation expresses the relation between Y for a system of size N, to Y for a system λ times larger. When the scale factor Z depends only on λ, i.e. Z(λ,N) = Z(λ), equation (2) can be solved uniquely to give the scale-invariant result of equation (1) with $Z(\lambda) = \lambda^{\beta}$. Scale-invariance implies that such a relationship—the ratio Y(λN)/Y(N)—is parameterized by a single dimensionless number β, usually referred to as the scaling exponent. The quantity Y(λN)/Y(N) is independent of the particular system size N but is dependent on the ratio between sizes λ. This behavior is what produces the linear relationship when logarithms are taken of both sides of equation (1), and the resulting straight-line on a log-log plot is the signature of a power law.

Recent research has pinpointed that cities can exhibit distinct types of scaling relationships across various urban phenomena or properties [11]. Sub-linear scaling (when the β exponents take a value of less than 1) parallels the allometric scaling laws observed in living organisms and represents the existence of economies of scale arising from an increase in efficiencies through the sharing of infrastructure; it is exhibited in electrical grids (through the length of electrical cables) and road systems (length of roads or amount of road surface) among other things. Super-linear scaling (when the β exponent is greater than 1) appears to be unique to social systems and is closely associated with the concept of network effects that lead to human ingenuity and creativity. Super-linear scaling has been identified in the number of new patents, inventors, R&D employment, total wages, etc. Linear scaling (when the β exponent is approximately equal to 1) signifies a proportional increase in urban phenomena/metrics with size.

The observation of scale invariance implies that the effects of increasing population size are general and can be observed by comparing any two cities, regardless of their size. If, for example, Y measures economic output, and two urban areas have population sizes of N and λN, respectively, scaling implies that the ratio of their outputs is a function of the proportion of their population sizes λ, but not of N. Scaling relations manifest an important empirical property: the phenomenon, repeats itself (albeit nontrivially) on different scales [26]. Such repetition points to possible underlying dynamical or stochastic processes generating and maintaining the same relationship among structural and functional variables over the range of the scale – typically many orders of magnitude [27]. A well-known example of a scaling relationship in the urban realm is "Zipf's Law", which states that a city's population decreases in inverse proportion to its rank among other cities within the same urban system [28], [29].

Population size and energy consumption in cities have often been analyzed through the concept of "urban metabolism". The concept of urban metabolism acknowledges that cities require a variety of inputs, among them energy, to maintain structure and remain functional [30]. Since its introduction in 1965, "urban metabolism" has become a widely used framework for understanding cities as both socio-economic and biophysical entities [31]–[36]. However, if CO_2 emissions can be interpreted as an indirect measure of urban energy use, the concept of urban metabolism invites a comparison with the biological realm. One of the most celebrated relationships in biology is the scaling relationship between metabolic rate and organismic mass. "Kleiber's law" states that for a vast array of organisms, metabolic rate scales to the ¾ power of the animal's mass [37]–[40]. That is, larger animals consume more energy than smaller ones but the rate at which energy is used increases less than proportionally to the increase in body size. Larger organisms are therefore more energy efficient than smaller ones. The analogy implicit in the widespread use of the concept of "urban metabolism" lends itself to a question: are larger urban areas more efficient (e.g. $\beta < 1$) than smaller ones with regards to CO_2 emissions?

Before proceeding to a discussion of the data and a presentation of results, we briefly address the use of level vs. per capita measures when examining a scaling relationship between two variables—as captured by Equation (1), specifically the usefulness of a per capita measure of CO_2 (such as CO_2 emissions per urban inhabitant) as compared to a measure of total CO_2 emissions for a population. When applied to urban metrics this presumes that urban characteristics scale linearly with city population size. If a scaling relationship exists between a variable Y and population, dividing Y by population introduces a nonlinearity into the per capita measure thereby reducing its accuracy [19], [41], [42]. Behind the choice of the most adequate dependent variable—total or per capita CO_2 emissions—lies a choice as to how to analytically approach cities: as extensive systems with constant size-independent densities (per capita quantities) or as non-extensive systems for which densities are non-intensive and thus highly variable [43].

Cities show extreme spatial and individual heterogeneity: individuals, households and businesses differ markedly with respect to their attributes and performance. There is no such thing as a representative business or average person inside the city. Furthermore, many of the properties of the basic constituting elements of a city depend on the size of the entire system. CO_2 emissions, as an extensive property, is accurately recorded in the aggregate but not in terms of the individual contributions. A scaling relationship is therefore a meaningful way of capturing how scale affects CO_2 emissions.

5.2 MATERIALS AND METHODS

We use CO_2 emissions data from Project Vulcan that quantifies U.S. fossil fuel carbon dioxide emissions at 10 km×10 km grid and at the scale of individual factories, power plants, roadways and neighborhoods on an hourly basis [44]. CO_2 emissions quantification utilizes datasets such as air quality emissions reporting, census data, highway vehicle use reports, energy use statistics, power plants emissions compliance reports, and econometric data [44], [45]. Furthermore, Vulcan includes significant process-level detail, dividing the emissions into 9 economic sectors and 23 fuel types [45]. We utilize the Vulcan data that is available at the level of counties for the years 1999 to 2008.

The U.S. spatial units of analysis are the 366 Metropolitan Statistical Areas (MSAs) and the 576 Micropolitan Areas, which together constitute the 942 urban 'core based statistical areas' (CBSAs) of the United States. An MSA is defined as an "urbanized area" (densely settled areas with a population of at least 50,000) comprised of a central county together with adjacent outlying counties having a high degree of social and economic integration with the central county as measured through commuting flows. The geographical boundaries of MSAs can thus be identified as the outer boundaries of the set of counties that comprise them. A Micropolitan Area is similarly defined but the urbanized area has a population of less than 50,000 but greater than 10,000. Note that the county definition for urban areas experienced very little change over the decade for which the data on carbon emissions is available. In 2010, 83.7% and 10% of the U.S. population resided in MSAs and micropolitan areas respectively; 6.3% lived outside of MSAs and micropolitan statistical areas [46].

CBSA definitions are independent of municipal or State governmental jurisdictions or boundaries; MSAs and Micropolitan Areas constitute in effect unified labor markets. The range of population sizes exhibited by Metropolitan and Micropolitan Areas goes from Tallulah, Louisiana, with 12,113 inhabitants in 2010, to the New York metropolitan area with a population of almost nineteen million. These varied places provide their inhabitants with a social experience recognizable as "urban." The U.S. Census—through its Office of Management and Budget (OMB) Bulletins—updates and revises delineations of metropolitan and micropolitan areas periodically. Our dataset thus includes all "urban" settlements of the U.S., which generate approximately 97% of the nation's economic output, house about 94% of the country's population and occupy less than 23% of its total land area.

We aggregate the total population of each county in the U.S into the MSA and micropolitan totals, using data from the Department of Commerce's Bureau of Economic Analysis (BEA). We also aggregate the total amount of CO_2 emissions (measured in millions of metric tones) allocated to each county by the Vulcan Project into MSA and Micropolitan Area totals based on the 2008 county delineations for metropolitan and micropolitan areas provided by the Census Bureau. We then construct a panel dataset for the period 1999–2008. Note that we aggregate all of the sources of CO_2 emissions because we are interested in the energetic aspect of urban life and not simply on any one component–it could be that the compact spatial form of cities is associated with gains in energy efficiencies but that these gains are offset by the increased consumption facilitated by higher productivity levels induced by larger urban agglomerations.

Following our emphasis on scaling effects, we hypothesize that urban CO_2 emissions are closely related to population size and that it scales according to a power-law relationship measured by.

$$Y_{i,t} = Y_0 N_{i,t}^{\beta} \tag{3}$$

where Y measures total CO_2 emissions, Y0 is a constant, N denotes population, β is the scaling exponent, and i and t index the urban area and year, respectively. This polynomial is a ubiquitous functional form commonly used in the natural and social sciences. Equation (3) acts as a baseline model and we let the data determine whether urban CO_2 emissions are adequately modeled with a power-law relationship.

5.3 RESULTS

We use a decade of data for each urban area and across all urban areas to estimate a panel for Equation (1) using a generalized least squared framework which corrects for AR(1) autocorrelation within panels and cross-sectional correlation and heteroskedasticity across panels [47]. Our 930 cross-sectional observations across 10 years provide a total of 9,330 observations. Taking the logarithms of both sides of Eq. 3 and suppressing the panel (i,t) notation, our model yields the following result:

$$\begin{aligned} \ln(CO_2) &= 2.35 + 0.933\ln(population), R^2 = 0.99 \\ &\quad (0.101)(0.008) \end{aligned} \tag{4}$$

The 95% confidence interval for the ln(population) coefficient in Eq. 4 is [.9164905,.9499573]. The coefficient is thus statistically different than 1. The scaling coefficient can be interpreted as elasticity, where a 1% increase in population size is associated with a nearly proportional increase in CO_2 emissions of 0.93%. The value in parentheses is the heteroskedasticity-corrected standard error. Note that the same model and specification, run only for the subsample of MSAs for the 10 years (leading to a total of 3630 observations) yields a ln(population) coefficient of 0.90 and the same level of R2. We also conduct cross-sectional OLS estimations for each of the ten years for which data is available, done with a correction for heteroskedasticity; these regressions yield scaling coefficients in the order of 0.93–0.95 (a remarkable stability across time) and R^2 values ranging from 0.67–0.76. Using only the subsample of MSAs, the OLS estimations for each of the ten years, correcting for heteroskedasticity, yield

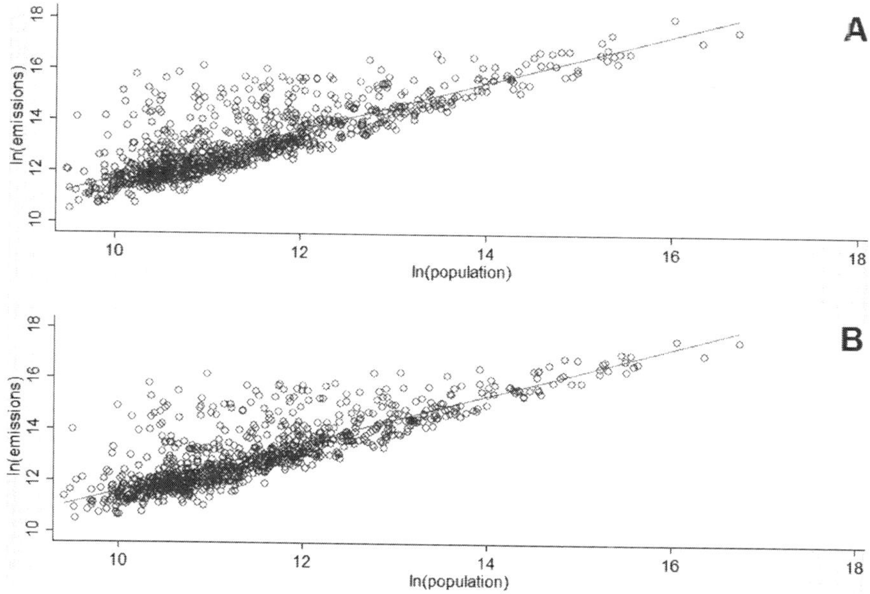

Figure 1. Cross-sectional log-log regressions for years (A) 1999 and (B) 2008.

scaling coefficients in the order of 0.91–0.92 and R2 values ranging from 0.67–0.68. Figure 1 plots the cross-sectional regression results for the full sample and the two endpoint years in our dataset.

Figure 2 plots the residuals from the full-sample cross-sectional regression for year 2008. Residuals range from a minimum value of −1.4 to a highest value of 3.9 but the vast majority range between [−1, 1]. Micropolitan areas produce the highest positive residuals and the highest negative residuals in our analysis, compared to MSAs.

Table 1 provides specific examples of the residuals ranking of the top 20 MSAS in the United States in year 2008. The biggest 20 MSAs in the U.S. span across a broad spectrum of the residuals ranking as shown in Figure 2. Cities like St. Louis and Minneapolis-St. Paul exhibit the highest positive deviations from the estimated scaling law in this subsample of the most populous MSAs. All MSAs that have positive residuals are

Table 1. The 20 most populous MSAs in 2008 ranked by their deviation from the scaling law.

Top-20 MSAs (population) in 2008	Residual Rank	Deviation from scaling law
St. Louis, MO-IL	125	Positive
Minneapolis-St. Paul-Bloomington, MN-WI	158	Positive
Atlanta-Sandy Springs-Marietta, GA	195	Positive
Chicago-Joliet-Naperville, IL-IN-WI	209	Positive
Detroit-Warren-Livonia, MI	232	Positive
Houston-Sugar Land-Baytown, TX	236	Positive
San Francisco-Oakland-Fremont, CA	244	Positive
Tampa-St. Petersburg-Clearwater, FL	283	Positive
Baltimore-Towson, MD	307	Positive
Washington-Arlington-Alexandria, DC-VA-MD-WV	315	Positive
Phoenix-Mesa-Glendale, AZ	356	Negative
Boston-Cambridge-Quincy, MA-NH	432	Negative
Dallas-Fort Worth-Arlington, TX	475	Negative
Riverside-San Bernardino-Ontario, CA	485	Negative
Philadelphia-Camden-Wilmington, PA-NJ-DE-MD	533	Negative
San Diego-Carlsbad-San Marcos, CA	568	Negative
New York-Northern New Jersey-Long Island, NY-NJ-PA	664	Negative
Miami-Fort Lauderdale-Pompano Beach, FL	673	Negative
Seattle-Tacoma-Bellevue, WA	684	Negative
Los Angeles-Long Beach-Santa Ana, CA	778	Negative

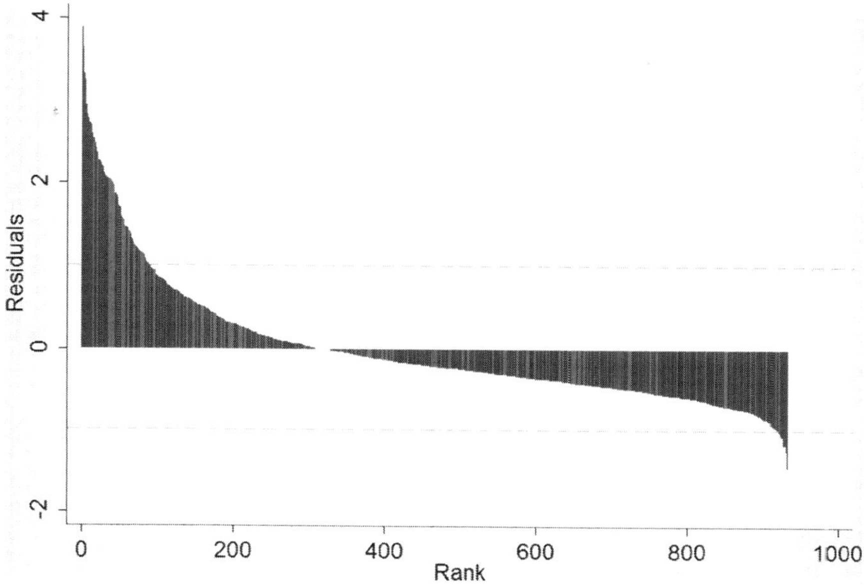

Figure 2. Ranking of residuals from the scaling regression for year 2008.

considered to be underperforming in terms of CO_2 emissions given their size. Cities like Los Angeles and Seattle exhibit the lowest negative deviations from the estimated scaling law in the subsample of MSAs. Cities with negative residuals are over-performing compared to the expectation based on their size.

Next, we enrich the relationship represented by Equation (4) with other important urban characteristics that may affect the energy consumption of urban areas: population density and residents' wealth. Studies show that certain population density thresholds (that vary by location) are required to support public transport. Additionally, higher population densities, coupled with higher employment densities, also enable mixed land use, which in turn is critical for non-motor vehicle transport [10], [47]. Here, we use population density as an indicator of land use mix and urban form. Population density reflects urban form which in turn affects how much the mobility of urban residents depends on the use of vehicles. An urban area's wealth is reflective of its economic composition and demographic characteristics, both of which may influence the intensity with which carbon-based fuels are used.

To control for the mediating effects of spatial form and wealth on the relationship between population size and urban energy use we add two independent variables to Equation (3), capturing the effects of urban wealth and population density. We define urban wealth as the per capita personal income (measured in current dollars). "Personal income" is the income received by individuals from all sources and is calculated as the sum of wage and salary disbursements, supplements to wages and salaries, personal transfers (such as social security payments), as well as proprietors', rental, dividend and interest income minus the contributions for government social insurance. "Per capita personal income" is obtained by dividing the total income accrued to the residents of an urban area by the area's population. Data on urban PCPI is reported by the Department of Commerce's Bureau of Economic Analysis (BEA).

We also create an urban population density measure that follows a population-weighted density definition [48], [49]. While a simple measure of density captures the ratio of urban population to total land area within the metropolitan boundaries, a population-weighted density measure resolves the problem of the non-uniform distribution of urban population within a city's administrative boundaries. Thus, our density measure uses the proportion of total metropolitan population found within a county as weights, and provides a more accurate variable of urban density as experienced by the average urban inhabitant. While our intent is to use this density measure to control for the effects of land use mix and urban form on CO_2 emissions it is important to note that the variable only imperfectly controls for the full range of potential urban form effects. Note that significant differences exist between the standard and the population-weighted density measures [46]. The New York MSA is almost twice as dense, while Phoenix is one and half times denser, using the population-weighted measure.

Including a measure for population density and per capita personal income as controls we obtain the following estimation results (Eq. 5) for a representative year (2008):

$$\ln(CO_2) = \underset{(1.31)}{1.685} + \underset{(0.029)}{1.028}\ln(population) - \underset{(0.037)}{0.172}\ln(density) + \underset{(0.133)}{0.364}\ln(pcpi), R^2 = 0.70 \quad (5)$$

Robust standard errors are reported in the parentheses. The 95% confidence interval for the ln(population) coefficient in Eq. 5 is [.971, 1.084]; the coefficient is thus statistically indistinguishable from 1. This finding is replicated across all years in our study, with coefficients ranging from 1.02–1.03. While the effect of population is now linear (rather than near linear as discussed above), the results indicate that an increase in population density decreases CO_2 emissions. In particular, in terms of elasticity, a 1% increase in our population-weighted density is associated with a 0.17% reduction in total CO_2 emissions, ceteris paribus. Across all years in our study, the estimated coefficients for ln(density) range from −0.172 to −0.149. The effect of density is always statistically significant across the years in our study. Our findings suggest that while emissions drop with density, the benefits from the added density (such as trip savings or shortening) are overshadowed by the effects of the size of the metropolitan area.

Furthermore, our analysis shows that, controlling for urban population size and average density, in 2008, differences in wealth have a small positive effect on CO_2 emissions – a 1% increase in personal income is associated with a 0.36% increase in total CO_2 emissions, ceteris paribus. Across the years in our study, we find that this small positive effect of personal income is typically not statistically significant at the 1% level (it becomes statistically significant only in the latter years of our timeframe, post-2005, and the estimate coefficient ranges between 0.26 and 0.36). This finding in partially conflicting with the general consensus on the effect of wealth on CO_2 emissions [23], [50]. Note that adding the density and wealth variables in the cross-sectional specification across all years does not improve the explanatory power of the models.

We also report the results utilizing the panel dataset and a generalized least squared framework which corrects for AR(1) autocorrelation within panels and cross-sectional correlation and heteroskedasticity across panels [47]; this approach though comes with a caveat: the personal income data is expressed in terms of nominal dollars (not real dollars), creating a challenge in the interpretation of the results from a panel regression (Eq. 6).

$$\underset{(0.27)(0.014)}{\ln(CO_2) = 3.9 + 1.057} \ln(population) - \underset{(0.017)}{0.163} \ln(density) - \underset{(0.026)}{0.22} \ln(pcpi), R^2 = 0.99 \quad (6)$$

While the population and density explanatory variables yield the expected magnitude and sign, a 1% increase in personal income is now associated with a 0.22% decrease in expected total CO_2 emissions, ceteris paribus.

5.4 DISCUSSION

Scaling is simply an emergent relationship between systemic size and emissions. Our results show that emissions in urban areas belong to a broader paradigm since every system needs to consume energy to maintain structure and order. The existence of approximate scaling phenomena for urban areas—documented using a variety of socio-economic metrics—is an indication that there are generic social mechanisms and properties of social systems at play across the entire urban system. Mechanisms such as networks and flows, nonlinearities and feedback loops integrate complex interactions among the individuals, households, firms, and institutions living, residing and operating in these spaces, leading to emergent phenomena such as scaling laws.

The near-linear relationship between population size and carbon emissions suggests that large urban areas in the U.S. are only slightly more emissions efficient than small ones. For each year in our sample, variation in population size across cities in the U.S. urban system explains approximately 70% of the variation of CO_2 emissions with density and wealth not adding explanatory power to the models. This figure does not change when considering only MSAs—that is excluding settlements with populations between 10,000 and 50,000 people. This leaves a substantial proportion of the variation to be explained in the cross-sectional data by factors other than total population, density and wealth. Overall, stated in terms of CO2 emissions savings, cities in the US do not exhibit economies of scale on average (as defined by the elasticity concept we estimate in this paper) since they scale almost linearly. We suggest that this can only be claimed "on average" because we are not testing for scaling across different population types (e.g. we do not examine a potentially deviating scaling relationship arising from population specializing in distinct industrial sectors). That is, while more substantial economies of scale may be present when a city grows in terms of service sector or "creative" professionals, no economies may be present when the same cities adds manufacturing jobs. Our finding represents the average effect in the specific ten year evolution of the U.S. urban system. Controlling for variation in population density and wealth in cities does not alter our findings.

The intuitive interpretation of the linear scaling finding can be explored first through the analogy urban metabolism. Our finding creates a paradox when one considers that in nature, as organisms grow in size they become more efficient (see discussion on Kleiber's Law above). A near-linear scaling in CO_2 emissions, and thus only marginal gains in efficiency, casts some doubt on the hypothesis that urban systems function similarly to biological ones. While the analogy between urban metabolism and biological metabolism has been questioned before [36], our analysis provides further evidence that the analogy may have empirical limits. We now know that cities exhibit characteristics that make the natural organism analogy difficult, such as the urban phenomena that produce super-linear scaling [11]. Still, a theoretical possibility that energy use scales sub-linearly but CO_2 emissions scale linearly; this would be the case if efficiencies in energy use where overshadowed by increased carbon intensiveness of the energy source mix that serves larger cities, the fossil fuel intensiveness of energy used in larger cities or the energy required to produce a unit of GDP in larger cities. Energy and emissions could scale differently because emissions are dependent on the type of energy used to generate final energy, the technology employed to use the energy, and the energy intensity of the economy [51].

We thus argue that an intuitive interpretation of the linear scaling finding requires an interpretation from economics, combined with an understanding of the nature of greenhouse gas emissions in the US. CO_2 emissions depend significantly on the carbon intensity of the energy source and the drivers of demand for fossil fuels. Several hypotheses can be made on the basis of a decomposition of factors that drive demand for fossil fuels in localized markets. Expecting a pattern of increased savings in CO_2 in larger urban agglomerations, a linear scaling of CO_2 emissions may signify that larger urban areas are lagging in their capacity to curb demand for fossil fuels proportionally to smaller urban areas. Or, it may be the case that residents in larger urban areas are not incentivized structurally (through urban form) or economically (through energy prices) to demand lower proportions of fossil fuels in their energy mix. Furthermore, although large urban areas are more innovative than smaller ones, they may lack capacity in steering eco-innovations towards their local markets for fossil fuels. These important hypotheses remain untested and need to be addressed in future research.

Notwithstanding, our results have important energy policy implications for a rapidly urbanizing planet since they reveal the importance of urban scale/size relative to factors such as population density and wealth. The research shows that policymakers need to renew their attention on issues of distributions of city sizes within national urban systems; we show that size trumps the effects of all other variables (such as population density and wealth) in explaining variation in CO_2 emissions. A focus on urban densities and wealth is still required, as these factors are critical for addressing various facets of global environmental change related to urban development. But as (new) world cities continue to grow, it is important that policymakers consider the CO_2 emission effects of urban size and contrast it to the effects of urban form, building materials and transportation network structure. While we expect that scaling laws characterize the structure and order of urban systems globally, whether our specific U.S. results hold for all typologies of cities is beyond the scope of this study [52], [53].

The issues associated with emissions and energy accounting methods highlight the limitations of assuming cities as "closed systems". The "closed system" perspective is in large part driven by the dominant conceptualization of a city through its narrow administrative boundaries – a definition of urban areas that drives data collection globally and dominates research practice surrounding urban phenomena. As we build our capacity to associate the increase of a city's size to effects that occur far away from a city's boundaries, we can overcome the data-specific challenge and adopt an "open system" perspective that could drastically alter our perspective on urban scaling. Through this new perspective, wealth, for example, may be found to be a more significant driver of total urban emissions; this is especially the case when considering emissions that occur in distal locations (or carbon sequestration capacity that is lost in distal places) but can be attributed to demand of goods and services that arises in specific urban areas [56]–[58].

Our "closed system" approach findings question the efficacy of using urban size as a climate change mitigation strategy. Our results show that, at least in the case of U.S. cities, there are no significant economies of scale with city size and CO_2 emissions. Therefore, cities and policies must consider other mitigation strategies that have been shown to have greater impacts on emissions than population size. Furthermore, considering the policy relevance of these findings, we claim that limited economies of scale with respect to carbon emissions should be viewed in conjunction to the build-up of additional evidence on urban scaling. Any strategic decision on city growth considering sustainability will have to carefully weigh the implications of urban scale on a variety of urban metrics (including innovation, crime, environmental indicators, etc.). Our results contribute to the larger picture of scaling relationships present in urban systems: given that larger cities "speed up" the process of wealth creation and innovation [11] and do not offer significant economies of scale in CO_2 emissions, a policy favoring larger city sizes may bring about carbon reductions primarily through technological advancements and eco-innovations.

REFERENCES

1. Seto KC, Sanchez-Rodriguez R, Fragkias M (2010) The new geography of contemporary urbanization and the environment, Annu Rev Environ Resour. 35: 167–194
2. GEA (2012) Global energy assessment – toward a sustainable future, Cambridge, UK: Cambridge University Press. 1888 p.
3. IEA (2012) World Energy Outlook 2012, Paris: International Energy Agency (IEA). 700 p.
4. UN (2012) World Urbanization Prospects, the 2011 Revision. New York, NY: United Nations, Department of Economic and Social Affairs, Population Division. 302 p.
5. Jones DW (1991) How urbanization affects energy use in developing countries. Energy Policy 19: 621–630

6. Parikh J, Shukla V (1995) Urbanization, energy use and greenhouse effects in economic development, Glob Environ Change. 5: 87–103
7. Puga D (2010) The magnitude and causes of agglomeration economies. J Urb Econ 50: 203–219
8. Dodman D (2009) Blaming cities for climate change? An analysis of urban greenhouse gas emissions inventories. Environ Urban 21: 185–201
9. NRC (2009) Driving and the built environment: the effects of compact development on motorized travel, energy use, and CO2 emissions. Washington, D.C.: Transportation Research Board, National Academies. 227 p.
10. Ewing R, Cervero R (2010) Travel and the built environment: a meta-analysis, J Am Plann Assoc. 76: 1–30
11. Bettencourt LMA, Lobo L, Helbing D, Kühnert C, West GB (2007a) Growth, innovation, scaling, and the pace of life in cities. PNAS 104: 7301–7306 [PMC free article] [PubMed]
12. Rosenthal SS, Strange WC (2004) Evidence on the nature and sources of agglomeration economies. In: Handbook of Regional and Urban Economics Henderson JV, Thisse JF, editors. 4: 2119–2171
13. Carlino GA, Chatterjee S, Hunt RM (2007) Urban density and the rate of invention. J Urb Econ 61: 389–419
14. Bettencourt LMA, Lobo J, Strumsky D (2007b) Invention in the city: increasing returns to patenting as a scaling function of metropolitan size. Research Policy 36: 107–120
15. Lobo J, Strumsky D (2008) Metropolitan patenting, inventor agglomeration and social networks: A tale of two effects. J Urb Econ 63: 871–884
16. Sedgley N, Elmslie B (2011) Do we still need cities?: evidence on rates of innovation from count data models of metropolitan statistical area patents. Am J Econ Sociol 70: 86–108 [PubMed]
17. Glaeser EL, Resseger MG (2010) The complementarity between cities and skills, Journal of Regional Science. 50: 221–244
18. Arbesman S, Christakis NA (2011) Scaling of Prosocial Behavior in Cities. Physica A 390: 2155–2159 [PMC free article] [PubMed]
19. Bettencourt LMA, Lobo J, Strumsky D, West GB (2010) Urban scaling and its deviations: revealing the structure of wealth, innovation and crime across cities. PLoS ONE, 5, e13541. [PMC free article] [PubMed]
20. Smil V (2008) Energy in nature and society: general energetics of complex systems, Cambridge: MIT Press. 496 p.
21. Payne JE (2010) Survey of the international evidence on the causal relationship between energy consumption and growth. J Econ Studies 37: 53–95
22. Lobo J, Strumsky D, Bettencourt LMA (2009) Metropolitan areas and CO2 emissions: large is beautiful. Working Paper Series: Martin Prosperity Research.
23. Kennedy C, Steinberger J, Gasson B, Hillman T, Havránek M, et al. (2009) Greenhouse gas emissions from global cities, Environ Sci and Technol. 43: 7297–7302 [PubMed]
24. Glaeser EL, Kahn ME (2010) The greenness of cities: carbon dioxide emissions and urban development, J Urb Econ. 67: 404–418

25. Lobo J, Bettencourt LMA, Strumsky D, West GB (2013) Urban Scaling and the Production Function for Cities. PLoS ONE 8(3): e58407.doi:10.1371/journal.pone.0058407 [PMC free article] [PubMed]
26. Barenblatt GI (2003) Scaling, Cambridge: Cambridge University Press. 188 p.
27. Sornette D (2004) Critical phenomena in natural sciences - chaos, fractals, self-organization and disorder: concepts and tools. 2nd Ed. Berlin: Springer Verlag. 528 p.
28. Zipf GK (1949) Human behaviour and the principle of least effort, Reading, MA, Addison-Wesley. 573 p.
29. Gabaix X, Ioannides YM (2004) The evolution of city size distributions. In: Handbook of Regional and Urban Economics Henderson JV, Thisse, JF, editors. 4: 2341–2378
30. Brown JH, Burnside WR, Davidson AD, DeLong, JR, Dunn WC, et al. (2011) Energetic limits to economic growth. BioScience 61: 19–26
31. Wolman A (1965) The metabolism of cities. Sci Am 213: 179–190 [PubMed]
32. Baccini P (1997) A city's metabolism: towards the sustainable development of urban systems. J Urb Technology 4: 27–39
33. Newman PWG (1999) Sustainability and cities: extending the metabolism model. Landsc Urban Plan 44: 219–226
34. Gandy M (2004) Rethinking turban metabolism: water, space and the modern city. City 8: 363–379
35. Kennedy C, Cuddihy J, Engel-Yan J (2007) The changing metabolism of cities. Journal of Industrial Ecology 11: 43–59
36. Golubiewski N (2012) Is there a metabolism of an urban system? An ecological critique. Ambio 41: 751–764 [PMC free article] [PubMed]
37. Kleiber M (1961) The fire of life: an introduction to animal energetics. New York, NY: Wiley. 454 p.
38. Bonner JT (2006) Why size matters. Princeton, NJ: Princeton University Press. 161 p.
39. West GB, Brown JH, Enquist BJ (1997) A general model for the origin of allometric scaling laws in biology. Science 276: 122–126 [PubMed]
40. Kolokotrones T, Savage V, Deeds EJ, Fontana W (2010) Curvature in metabolic scaling. Nature 464: 753–756 [PubMed]
41. Uslaner EM (1976) The pitfalls of per capita. Am J Pol Sci 20: 125–133
42. Katz JS (2006) Indicators for complex innovation systems. Research Policy 35: 893–909
43. Bettencourt LMA (2012) The origins of scaling in cities, Santa Fe Institute Working Paper, 12–09–014.
44. Gurney KR, Mendoza DL, Zhou YY, Fischer ML, Miller CC, et al. (2009) High resolution fossil fuel combustion CO2 emission fluxes for the United States. Environ Sci Technol 43: 5535–5541 [PubMed]
45. Gurney KR, Chen YH, Maki T, Kawa SR, Andrews A, et al. (2005) Sensitivity of atmospheric CO2 inversion to seasonal and interannual variations in fossil fuel emissions. J Geophys Res. 110: 10308–10321
46. Wilson G, Plane DA, Mackun PJ, Fischetti TR, Goworowska, etal. (2012) Patterns of metropolitan and micropolitan population change: 2000–2010, 2010

Census Special Reports, U.S. Department of Commerce, Economics and Statistics Administration.
47. Beck N, Katz JN (1995) What to do (and not to do) with time-series cross-section data. Am Pol Sci Rev 89: 634–647
48. Glaeser EL, Kahn ME (2004) Sprawl and urban growth. In: Handbook of Regional and Urban Economics Henderson JV, Thisse JF, editors. 4: 2481–2527
49. Rappaport J (2008) A productivity model of city crowdedness. J Urb Econ. 63: 715–722
50. Kahn ME (2006) Green cities: urban growth and the environment. Washington, D.C.: Brookings Institution Press, 170 p.
51. Kaya Y (1990) Impact of Carbon Dioxide Emission Control on GNP Growth: Interpretation of Proposed Scenarios." Paper presented to Intergovernmental Panel on Climate Change Energy and Industry Sub-group, Response Strategies Working Group. Paris.
52. Angel S, Parent J, Civco DL, Blei A, Potere D (2011) The dimensions of global urban expansion: estimates and projections for all countries, 2000–2050. Prog Plann 75: 53–107
53. Seto KC, Fragkias M, Güneralp B, Reilly MK (2011) A meta-analysis of global urban land expansion. PLoS ONE 6(8): e23777. [PMC free article] [PubMed]
54. Wackernagel M, Rees W (1996) Our Ecological Footprint: Reducing Human Impact on the Earth. Gabriola Island, BC: New Society Publishers. 160 p.
55. Burger JR, Allen CD, Brown JH, Burnside WR, Davidson AD, et al. (2012) The macroecology of sustainability. PLoS Biol 10: e1001345.doi:10.1371/journal.pbio.1001345 [PMC free article] [PubMed]
56. Seto KC, Reenberg A, Boone CG, Fragkias M, Haase D, et al. . (2012) Urban land teleconnections and sustainability, Proceedings of the National Academy of Sciences. [PMC free article] [PubMed]
57. DeFries RS, Rudel T, Uriarte M, Hansen M (2010) Deforestation driven by urban population growth and agricultural trade in the twenty-first century, Nature Geoscience 3, 178–181.
58. Weinzettel J, Hertwich EG, Peters GP, Steen-Olsen K, Galli A (2013) Affluence drives the global displacement of land use, Global Env Change. 23: 433–438.

There are supplemental files that are not available in this version of the article. To view this additional information, please use the citation on the first page of this chapter.

PART III

SUSTAINABLE LAND USE AND INFRASTRUCTURE

CHAPTER 6

Maintaining Climate Change Experiments: Urban Political Ecology and the Everyday Reconfiguration of Urban Infrastructure

VANESA CASTÁN BROTO AND HARRIET BULKELEY

6.1 INTRODUCTION

Addressing climate change in the city requires an engagement with the urban infrastructure systems that mediate the production and consumption of energy and greenhouse gas emissions, and which shape urban vulnerability to the impacts of climate change (Monstadt 2009; Hodson and Marvin 2011). While analyses of urban climate change responses have typically focused on the policy sphere and the development of plans and strategies to address the issue, acknowledging that urban infrastructure, understood as the sociotechnical matrix that facilitates the provision of services needed for urban life, is critical to any urban response opens up the possibilities for looking elsewhere in the city at how climate change is being addressed. Seen in this way, the plethora of seemingly one-off or ad hoc interventions — including behaviour-changing policies, development of new energy sources and new systems of water delivery, transport and waste collection — are part of the growing urban response to the challenges of climate change.

Some such interventions, which we term 'climate change experiments' (Castán Broto and Bulkeley, 2013), are purposively designed to trial the social and technical experience of responding to climate change, put new materials, technologies and social actions to the test, or develop knowledge within the city to respond to climate change. In the context of uncertainty, urban climate change interventions become experiments, or projects in the making, in the sense that they are used to explore uncharted policy territories, to either learn or open up new forms of intervention, and often without consideration of their unintended consequences. Understanding how such experiments take effect becomes a critical means to explain urban responses to climate change, because 'experiments serve to create new forms of political space within the city, as public and private authority blur, and are primarily enacted through forms of technical intervention in infrastructure networks, drawing attention to the importance of such sites in urban climate politics' (Bulkeley and Castán Broto, 2013). Moreover, as specific interventions in infrastructure networks, experiments challenge current understandings of urban responses to climate change and, more fundamentally, the conceptualization of the dynamics of such networks.

In this article, we interrogate one particular aspect of this process — maintaining experiments — more closely. Maintenance emerges as a key aspect of urban climate change experimentation, as a means to manage its inherent uncertainty. To theorize experiment maintenance we adopt the following strategy: in the first part of the article, we draw on literatures from urban political ecology concerned with the processes of metabolic circulation (Gandy, 2004; Kaika, 2005; Keil, 2005b; Heynen et al., 2006; Swyngedouw, 2006; Monstadt, 2009) and from urban geography on the process of 'repair and maintenance' (Thrift, 2005; Graham and Thrift, 2007; Strebel, 2011) to suggest that this is at once a strategic, political process, and one shaped through the experiences and practices of everyday urbanism. Conceptualizing maintenance in this dual fashion allows us to examine the ways in which climate change experiments become embedded within and serve to reconfigure forms of urban metabolism. In the second part of the article, we adopt a comparative approach as a critical strategy for creative learning (Robinson, 2006; McFarlane, 2010) about the materialization of climate change discourses in different urban settings. Urban climate change experimentation involves the transposition of best practice examples or models across cities (for examples, see Bulkeley, 2010). As 'policies in motion' (Ward, 2006), climate change experiments adjust to and reconfigure the settings in which they are deployed, in this case, two case studies in the housing sector in Bangalore (India) and Monterrey (Mexico).

The theoretical framework is mobilized to understand how urban climate change experiments are maintained or otherwise within the infrastructure fabric of the city. We find that this is not achieved through rendering infrastructures as 'invisible', as much of the writing on urban infrastructure suggests (e.g. Star, 1999), but rather requires an explicit and ongoing work of maintenance to integrate and assert the experiment within particular urban contexts. These processes we describe as 'metabolic adjustment', involving the connection and disconnection of urban flows, and 'upkeep', consisting of multiple urban practices leading to new infrastructure configurations. They constitute a key aspect through which the innovative character of experiments is sustained and reworked, while also ensuring that they become accepted. We argue that such processes are critical to understand the effect of experiments within the context of urban development in which they emerge.

6.2 URBAN INFRASTRUCTURES AND THE MAINTENANCE OF URBAN LIFE

Scholarship in urban political ecology has drawn attention to the ways in which the conditions of urban life are structurally produced and maintained through the continual circulation of capital, resources and nature (e.g. Swyngedouw and Heynen, 2003; Keil, 2005a; Heynen et al., 2006; Monstadt, 2009; Cooke and Lewis, 2010). Flows of 'metabolic circulation' are central to the realization of the urban condition because they enable the exploitation of natural resources and, in turn, produce the 'enabling and disabling social and environmental conditions' of the city (Monstadt,

2009: 10; see also Heynen et al., 2006; Swyngedouw, 2006). Urban political ecology analyses demonstrate that the transformation of ecological systems into the city is vital because 'urban politics has fundamentally to be about modes of transformation of nature related dialectically to modes of self-realization of a particular form of human nature' (Harvey, 1996: 435). This transformation is orchestrated through policies and interventions that require the ordering of urban natures, to separate 'good' and 'bad' natures according to socially constructed understandings of what the city ought to be (Desfor and Keil, 2004; Kaika, 2005).

Infrastructure networks are seen as central to these processes of transformation, as the 'functional lattice' through which material flows and flows of power are mediated (Gandy, 2004). Infrastructure, as a socio-technical configuration, is produced by these circulations and the means through which urban metabolism takes place. Circulation is inherent to the processes whereby wider circuits of capital and politics structure urban metabolisms and confer stability to the city. Through these processes of circulation and flow the apparent permanence and fixity of urban forms and governance is conferred (Harvey, 1996) and networks are rendered invisible (Star, 1999). This obduracy can be understood as caused by the embeddedness of different elements in closely interrelated assemblages of socially constructed and technically produced elements (Hommels, 2005, 2008). At the same time, these processes are dynamic, emergent and unruly, involving both the continual reproduction of urban conditions and an inherent unpredictability. Maintaining urban infrastructure can be seen in this view as a broadly structural process of achieving or containing particular forms of metabolic circulation orchestrated by forms of connection, disconnection and fragmentation. Through this process infrastructure networks are simultaneously challenged and invested with obduracy to ensure the dual production of nature and capital circulations within the city. In this manner, maintenance becomes critical to the reproduction of the material/infrastructure in everyday life, and to its continuing structuring effects in the city.

As specific interventions in infrastructure networks, and in line with the broad literature on urban political ecology, experiments require a reworking of the flows of power, resources and materials through which infrastructure systems are sustained. Achieving this intervention, we argue, requires a process of 'metabolic adjustment' through which experiments may be embedded within particular circulations and reconfigure the infrastructural 'lattice' of the city. As a form of metabolic adjustment, maintenance processes sustain and redirect metabolic flows and become a critical means through which the reconfiguration of infrastructure networks takes place. Maintaining experiments can therefore be regarded as a process in which new rationales of nature and space production come to the fore.

However, maintenance is a structural process concerned with broad circuits and systems. As Graham and Thrift (2007) argued, everyday practices in the city also play a critical role in the (often hidden) processes of repair and maintenance, which constitute the 'urban technological unconscious which helps to keep cities as predictable objects in which things turn up as they are meant to, regularly and predictably' (Thrift, 2005: 136). Graham and Thrift (2007: 4) position these processes as 'reality's bridge' between the 'ready-to-hand', taken-for-granted world of practical action and the visible disruption to the everyday life of failed things. Here, maintenance is understood as the practical, microscale means through which urban networks are kept in 'good repair', and in turn, through which particular sociotechnical configurations are sustained. This 'carpet of ongoing maintenance and repair' present in every city is characterized by five qualities: its reliance on the power of things; the pluricultural character of things, whereby 'they have come to play a role in the everyday life of almost everyone'; the growing demand for maintenance and repair created by the growing number of things and their complex material composition; the imperative created by the growing reliance on infrastructure networks for urban life to ensure decay and disruption is minimal; and increasing difficulty to 'define what the "thing" is that is being maintained and repaired' between objects, social relations and ways of urban living (Graham and Thrift 2007: 3–4).

This work specifically points to the range of mundane and intricate practices involved in maintenance, from street cleaning to the relaying of gas pipes. Everyday life is orchestrated through and around these activities, so that 'modern urban dwellers are surrounded by the hum of continuous repair and maintenance' (Thrift 2005: 136). In these readings, repair and maintenance are positioned, on the one hand, as a means through which the elemental decay of urban life is held at bay, constituting a critical part of urban (political) economy, and we would add, ecology; and, on the other hand, such processes are also regarded as central to the creation of new forms of urban life, as 'a vital source of variation, improvisation and innovation' (Graham and Thrift, 2007: 6). In this view, experimentation emerges in the process of maintenance for: 'when things break down, new solutions may be invented ... there is some evidence to suggest that this kind of piece-by-piece adaptation is a leading cause of innovation, acting as a continuous feedback loop of experimentation which, through many small increments in practical knowledge, can produce large changes. Seen in this light, "maintenance is learning"' (Brand, 1994: 127, quoted in Graham and Thrift, 2007: 5). Graham and Thrift demonstrate the ways in which maintenance is critical not only in reproducing forms of urban obduracy but also in facilitating urban transformations. Developing this argument, we suggest, entails understanding that the maintenance of change[1] is as significant as that of stasis or decay, and in itself creates a powerful force of urban permanence/immanence.

Graham and Thrift's seminal work on repair and maintenance provides a critical means through which to understand the mundane and practical ways in which the upkeep of infrastructure networks serves to continually reproduce the urban landscape. Yet, we find that maintenance is not limited to the upkeep of existing assemblages, but is central to the work of experimentation itself. This is because maintenance is about sustaining the existence of current sociotechnical relations or of keeping things in 'good repair' (anticipating failure, disruption or decay), and it is a means through which experiments are offered support and affirmation as they become normalized within particular urban contexts. As a set of mundane practices, maintenance can therefore be regarded as a means through which experiments are sustained and gain 'momentum' in the city. In other words, 'maintenance is not only an event through which relational assemblages of human and nonhumans are brought into view, but a course of action in which [in this case] the high-rise [building] gains momentum' as a 'living' entity (Strebel, 2011: 248). Furthermore, while Graham and Thrift focus on the micropractices and everyday nature of maintenance, drawing on the urban political ecology perspective above, we also suggest that this process is central to the reproduction of power and discourses that enable certain forms of metabolic circulation to become viewed as normal and embedded within the city.

Reading across these literatures from urban political ecology and urban geography, maintenance emerges as a set of processes that are, on the one hand, concerned with the strategic and structural dynamics of sustaining and reworking particular forms of metabolic circulation and, on the other, undertaken through a range of mundane and everyday practices (Table 1). As disruptive interventions within the 'infrastructural lattice' of particular urban contexts, we suggested that climate change experiments give rise to maintenance practices that reveal their critical role in the dynamic transformation of urban landscapes in response, in this case, to the climate change problematic. We identify here the potential for 'metabolic adjustment', as the experimental intervention reconfigures network flows and the micropractices involved in the 'upkeep' of experiments. Upkeep and metabolic adjustment processes have a dual function in the sense that they are simultaneously directed towards ensuring the permanence of the experiment within the given infrastructure context, and asserting its potential for provoking change. Thus, as demonstrated below, we find that maintenance is required to sustain the novelty of experiments and to normalize them within particular urban contexts.

6.3 MAINTAINING CLIMATE CHANGE EXPERIMENTS IN MEXICO AND INDIA

Climate change experiments, as explicit interventions in urban infrastructure systems that seek to reconfigure sociotechnical relations around climate change, provide a critical and intriguing site through which to further explore how forms of maintenance contribute to the reproduction and reconfiguration of urban landscapes. We selected, from a database of 627 urban climate change experiments, two cases on housing that were recognized as having a systemic impact in the city in which they occur.

The first case study is a gated development of 91 eco-homes for high-income professionals in the urban periphery of Bangalore (India) called Towards Zero-Carbon Development (T-Zed), led by a private developer called Biodiversity Conservation India Ltd (BCIL). The second case

Table 1. Representation of maintenance as a dual process of metabolic adjustment and upkeep.

		Dual function	
		Permanence	**Change**
Form of maintenance	Metabolic adjustment	Embedding the experiment within the existing urban metabolic circulations	Resignifying the experiment as a symbol of structural reconfigurations
	Upkeep	Normalizing the experiment in everyday practices	Asserting the experiment as a process in the making

study is a social-housing development of 51 houses in Monterrey (Mexico) called Vivienda de Diseño Ambiental (ViDA), led by the state-based housing institute (Instituto de la Vivienda de Nuevo Leon) and supported by the National Housing Commission (CONAVI) with the collaboration of Infonavit, a federal institute that provides mortgages for workers' housing. Fieldwork was carried out during 3 weeks in Bangalore (March 2010) and Monterrey (April–May 2010). The fieldwork consisted of visits to the sites and interviews with key actors who had intervened in the development directly or indirectly. There were 47 interviews in Bangalore and 31 in Monterrey. In Bangalore interviewees included key actors in T-Zed (14 interviews, 1 informal), residents (6 interviews) and other actors influencing local climate change governance, including NGOs, public sector officials and academics (33 interviews, 5 informal). In Monterrey interviewees included key actors in ViDA (11 interviews, 4 informal), residents (7 interviews, 3 informal) and other actors influencing local climate change governance, including NGOs, public sector officials and academics (13 interviews, 1 informal). Formal interviews were conducted in a formal setting, with an interview guide, and they were taped, transcribed and coded using a predetermined coding guide. In some cases (indicated above) only informal meetings were possible due to time or personal constraints. The analysis also examined a range of written sources including promotional materials, government press releases, newspaper articles, discussion articles in architectural blogs and forums and academic presentations. Table 2 summarizes some of the main characteristics of each development. In both cases, the housing developments challenge established sociotechnical arrangements within the housing system in an effort to address climate change.

Whereas, broadly, T-Zed and ViDA attempted to develop low-carbon models of urban development in their cities, their objectives could not be more different. T-Zed focuses on creating alternative markets for green developments while ViDA attempts the seamless integration of low-carbon development principles in an existing and well developed social housing industry. Their comparison sheds some light on the mutual relationship between experimentation and urban maintenance. In line with the theoretical discussion above, and following the analysis summarized in Table 1, we identify maintenance as consisting of two related components. First, we focus on the ways in which the experimental quality of these infrastructural interventions requires forms of everyday upkeep. Second, we articulate how maintenance as 'metabolic adjustment' takes place to redirect urban circulations around and through the experiment as an exceptional space, and to embed them in the specific sociotechnical arrangements of the city and neighbourhood. We argue that this form of maintenance involves the need to preserve the innovative character of the experiment and a process of normalizing the experiment so that it can be integrated within existing urban practices. This may require reframing the experiment to adapt to the specific discursive mechanisms seeking its reproduction at other scales. Through each form of maintenance, we find that there is a tension between the need for standardization of the experiment or experiment components, and the need to re-experiment the intervention to highlight its exceptionality and further its momentum.

6.4 THE EVERYDAY UPKEEP OF EXPERIMENTS: NOVELTY AND NORMALIZATION

Upkeep is a form of maintenance, which occurs in everyday practices. How upkeep occurs is tied to the design of the experiment and to the constellation of actors intervening and the urban infrastructure regime in

Table 2. Main characteristics of experiments.

Name of experiment	T-Zed	ViDA
Experiment led by	Biodiversity Conservation India Ltd (BCIL)	Instituto de la Vivienda de Nuevo León
Lead organization	Private	Public
Assembling resources	Commoditization/product making	Public provision of social housing
Infrastructure provision	Developer	Developer
Completion date	2007	2008
Main innovations	Carrying capacity design, traditional construction techniques, biodigester, community living	Design adapted to climate: orientation, ventilation, energy-efficient appliances

which it is deployed. In T-Zed, BCIL see themselves as 'helping people to get some sense in the decisions they make' (interview T-Zed D3[2]), that is, to help high-income middle-class professionals to combine green aspirations (global and local) with housing aspirations (big houses with modern commodities and secure supply of energy, water and waste). In ViDA, the project was made possible by promoters from the Instituto de la Vivienda de Nuevo Leon and CONAVI who kept in mind not the interests of future residents (who were thought to benefit indirectly from the project) but those of developers, who should be able to implement eco-technologies without increasing their costs.

Regarding urban infrastructure regimes, the landscape of Bangalore is one of fragmentation where multiple services for service provision co-exist. In terms of water, waste and energy, the city is provided through various governance arrangements which combine public provision in networked areas (mostly old quarters of the city) and private arrangements for high-income classes in new developments and for the poorest sectors of society which live in new layouts outside the main networks or in informal settlements (e.g. Ghosh, 2005; Ranganathan et al., 2009). Differences in governance arrangements of service provision have led to a spatial fragmentation of services — a splintering — whereby upcoming areas are able to securitize privately the provision of water and other services in gated communities (see Graham and Marvin, 2001). In Monterrey, permits for development are given under the condition that developers lay down the corresponding infrastructure according to municipal regulations (although regulations vary in each municipal term, leading to coordination problems). Municipal authorities take up responsibility for service provision after the development is complete, but they cannot always meet these because of challenges related to rapid growth and resource availability, particularly in terms of water provision and availability of land for construction with continuous struggles over the development pressures on the natural resources of the city.

The result is two differentiated forms of maintenance as upkeep. In T-Zed, we find a privatized form of maintenance, customer oriented, whereby BCIL is in charge of the everyday management of the site. According to the onsite manager, this is because the site requires 'special care that only BCIL can provide' (interview T-Zed D1). However, maintenance is also regarded as a failure to deliver the experiment in time: 'Construction was still going on two months back' (interview T-Zed R2), explained one resident in March 2010, having lived for 3 years on the compound. For some, this signifies a prioritization of environmental protection aspirations over delivery: 'because things are not finished … there are like levels and there are rough edges where you can hurt yourself … it seemed to me: "okay, you are trying to do all the environmental stuff but you are not really taking care of the people who are here" ' (interview T-Zed R4). Upkeep practices, here, are directed towards re-experimentation rather than normalization.

In ViDA, in contrast, the project was constructed and delivered in time. In the context of security of demand from workers guaranteed by Infonavit, developers are able to design highly standardized house types in which the use of materials is typified to minimize the costs of integration and transformation of materials available from suppliers. Thus, the architects from the Instituto de la Vivienda developed a design, which integrated bioclimatic construction principles while minimizing deviations from existing types. A representative of the Instituto de la Vivienda explained that they 'took a typical house prototype, broke that into pieces and put it together so that it had some [green] features' (interview ViDA P1). The architect responsible for the project explained that the pressure came from developers who told them 'not to move anything ... so we looked at the window, for example, not to change it but just to consider its orientation ... because when you make a project you need to understand the developers' (interview ViDA C1). Bioclimatic modifications had an additional cost, which was covered by the Instituto de la Vivienda providing land at a reduced price to facilitate the developers' financing. Overall, the houses were built according to plan, without challenging any standard procedures within the housing industry.

On completion, ViDA's houses were delivered to residents who would have to take care of the maintenance of each independent unit. With respect to the houses, residents explained that, beyond a few months' guarantee, the developer breaks any relationship with the development on completion. This is a common feature of housing developments under Infonavit. 'At this moment, we are not able to follow-up developments once they are completed', lamented an Infonavit representative (interview ViDA P3). As in other developments, a representative of the municipal department for public services explained that 'developers provide the pavement, street furniture, street lighting and other infrastructures [costs are included in the price of the house] and we confirm that they meet our specifications before taking the responsibility for the maintenance of services' (interview ViDA P2). However, as a representative of the local community explains, the community receives only limited support and it needs to take charge to ensure provision actually occurs. For example, a resident explained that in their relation with municipal services personnel 'the community needs to mark them the way and you have to be asking them continually whether they come or not' (interview ViDA R2). Ultimately, the maintenance of the development (not only of individual houses but also of common spaces such as the pavements and the park) is left in the hands of residents.

These cases demonstrate two fundamentally different models of upkeep. The T-Zed model relates to the involvement of the developer and the splintering urban configuration, which promotes private and customer-led patterns of service provision met collectively by the developer. In contrast, in ViDA, upkeep needs are determined by the developers' demands within the well-established conglomerate of the construction industry, finance and government, which effectively assumes maintenance to be outsourced to the municipality and the residents. Whether maintenance is regarded as a continuation of the experiment-making process (such as in T-Zed) or as a finished product (such as in ViDA) determines the upkeep protocols, regardless of residents' concerns (which in both cases complained about certain standard and operations, as explained below).

These different modes of governance also create material differences in the life of the experiment, in terms of individual dwellings, common spaces and the development as a whole. T-Zed, despite the signs of ongoing construction work (the rubble on one of the sides of the entrance; the sounds of construction work going on in some of the flats which had to be refitted; the barren piece of land which cannot be landscaped because its ownership is disputed), is perceived as a coherent whole. The house units in T-Zed are connected by gangways and gardens. Common units (a restaurant, pool, paddle, gardens and activity rooms) are extensions of collective spaces shared between houses, rather than additions. In ViDA, in contrast, there is no apparent connectivity between the houses. Residents explained that they felt they cannot use the collective areas freely (including front gardens, pavements and playground areas) because of safety fears. They also said that, apart from isolated efforts from some groups of residents, nobody maintains these common areas.

ViDA architects highlighted that, as an experimental project, some features of the project did not work as they expected. However, residents do not express concerns about the maintenance of individual households (although they may have adapted different elements to suit their needs). They compare their houses with similar alternatives, pointing at differences in terms of size and location. The 'bioclimatic' character of the houses is somehow eroded. Even the community elected representative (the 'Juez'), when prompted, said that if the house was bioclimatic 'that must have been because of the solar panel used to light the three bulbs outside' (interview ViDA R3), a feature which was incorporated only later in the design. Orientation, ventilation and other features of the house concern residents when they feel their negative consequences (for example, when strong winds cause dusting within the houses). Attempts to re-experiment the development, through a survey in 2010 about how residents were using the houses and parallel education campaigns, appear to have a minimal impact on the material operation of the housing development and on local residents' views.

T-Zed residents also compare the standards in their houses with those of developments for similar income bands. From this comparison, residents infer that the incorporation of multiple design and technological innovations have resulted in unexpected additional maintenance issues. Roof gardens leaked, unconventional construction materials did not provide the services expected, natural stones for tailing deteriorated faster than predicted: the whole development is a life project, still in the making. A BCIL customer-service worker explains some of these problems: 'initially we had dampness … in the walls … the customers had painted them in their own way and then we found bubbles in the paint, because the paint was not absorbed properly; then plumbing, then power problems, but [these are] normal maintenance issues, there is not anything specific for T-Zed' (interview T-Zed D6). What for BCIL are 'normal maintenance issues' for some residents is a cause for dissatisfaction and concern. As Graham and Thrift (2007: 4) argue 'accidents that stem from so many breakdowns are not aberrant but are a part of the thing itself'. Experimentation is inherently more prone to decay. Both in ViDA and T-Zed, maintenance is closely related to the power of things. However, while in ViDA, maintenance becomes embedded in residents' everyday practices of material production, in T-Zed normalization is led by the developer and maintenance becomes a process of expectations management.

The result is that BCIL's presence in T-Zed is needed to respond to the continuous demands of residents. In some cases, this has led to a renegotiation of the original terms, including refitting and alternative solutions for unsatisfactory technologies. In other cases, it has prompted a relaxation of BCIL's demands on residents about what appliances are allowed in the compound. For example, BCIL envisaged that, to reduce the consumption of energy, flats and houses would be provided with customized appliances for air conditioning and refrigeration. The fridges now pile in BCIL offices, as residents rejected the models (some found them too small and others simply could not get them to work) and there are ongoing discussions about whether the fridges were ever operative. After some disputes, residents are now 'free' to buy their own cooling appliances. Refrigeration was described by a consultant as one of T-Zed's major failures, and the system has been abandoned in new BCIL projects. Overall, BCIL confronts in T-Zed a proliferation of things being maintained and imperative demands for BCIL to provide immediate responses to each crisis. Even aesthetic issues provoke questions regarding the residents' lifestyles and exhibit, as Graham and Thrift (2007) highlight, 'an existential quality'.

These two contrasting models of upkeep, in relation to the actors involved in the experiment, reveal some of the fundamental features of maintenance. In both cases, the experiment has a life beyond assemblage, so that its elements need to be adjusted to deal with multiple changing relations through scrubbing, adapting, cleaning and reinventing as maintenance demands proliferate. This process will determine the eventual reassertion of the experimental character of the initiative. The proactive role of BCIL in T-Zed confers purpose to the experiment and reasserts innovations at the expense of compromising its normalization: continuous re-experimentation becomes a source of anxiety among residents. In ViDA, the lack of purposive maintenance gives way to conventional forms of repair and as a result becomes integrated in the landscape so that its potential ebbs away, despite architects' attempts at re-experimentation through surveys and education campaigns.

6.5 METABOLIC ADJUSTMENT: EMBEDDING AND REFRAMING EXPERIMENTS WITHIN URBAN FLOWS

Metabolic adjustment refers to a series of discursive and spatial processes that are needed to adjust the experiment to the conditions of urban development in which it emerges, while resignifying the experiment as a symbol of the intended structural configurations. Again, we find this is a dual process, which attempts to gain stability and assert the exceptionality of the experiment within the specific conditions in which it emerges. The comparative analysis of the cases of T-Zed and ViDA suggests that permanence and change are embedded in the experiment design but require maintenance in terms of asserting such designs through time and space. The key aspect common to both experiments is their emphasis on resource securitization (securing water resources, energy but also land and space), a feature that shapes how they unfold over time.

The most fundamental difference between T-Zed and ViDA is that they are targeted at different consumer groups: T-Zed is a gated compound for high-income professionals while ViDA is a social housing project for salaried workers. Each one represents the dominant trend in their respective cities. Following the unprecedented growth during the last two decades in Bangalore, new developments have been constructed to host big complexes of offices and residences for a rapidly growing class of middle and high-income professionals. In Monterrey, urban sprawl has occurred in a polycentric manner, with massive social housing developments organized in 'villages' and often centred on a particular industry, as is the case of the municipal term of Escobedo, in the conurbated area of Monterrey, where ViDA is located. What is common in both cases is that low-carbon experimentation in housing fails to challenge the accelerated process of urban sprawl. While T-Zed is oriented towards selling a suburban dream house to those who can afford it, ViDA is an example of how social housing contributes to the process of urban expansion in Mexican cities.

Moreover, neither example challenges fundamentally the housing industry or its drivers. In Bangalore, the construction industry is lured by the demands for space and comfort of the professional classes. Being mobile and cosmopolitan, this new class of customers demands a standard of service — from air conditioning to swimming pools — which is alien to the traditional city (Nair, 2000). While T-Zed designers advocate the use of local materials and sustainable technologies, they conform to imported models of suburban development. T-Zed is not different from nearby developments in that it seeks to enrol high-income customers with a premium product — characterized for being 'green' — in a gated community alongside one of the industry corridors in the outskirts of Bangalore. In Monterrey, like in the whole of Mexico, most social housing is financed by Infonavit, a public–private financial institution that has given more than 5 million mortgages to salaried workers since its creation in 1972. Infonavit is associated with a strong industry of housing development described by a local architect in an interview as 'a goose that lays homes' (interview ViDA, C1). This industry (consisting of a multitude of big and small private developers) has developed house types to make house units as cheap as possible, so that they can meet the requirements of social housing mortgages at Infonavit. The result is the perpetuation of repetitive models of construction in cheap land and an aversion to innovation, which has led to the exponential rise of urban sprawl in most Mexican cities. The integration of both experiments in the dynamics of urban development —in terms of form and design — exemplifies the permanence function of maintenance already embedded in the experiment's design.

However, this is not a smooth process, precisely because, as highlighted in the urban political ecology literature, the integration of the experiment with its surrounding environment depends on the infrastructure connections with the rest of the city. In T-Zed, the idea of a self-sufficient home, advanced here under a carbon reduction flag, contributes, on one hand, to create a luxury development in the context of lack of infrastructure for that type of service provision and, on the other, to securitize resources in the light of a future resource crisis that is affecting the whole city, particularly regarding water and energy. Self-sufficiency is also regarded as a benefit to the whole city: 'we are saving the cost of the entire society by lowering the load of what we are taking in electricity from the grid and by not connecting to the municipal sources' (interview T-Zed D2). This discourse emphasizes the type of private service provision characteristic of this kind of developments. Ultimately, private provision is here a strategy for resource securitization (Hodson and Marvin, 2009), which eventually leads to the splintering of infrastructures, such as in this case, water (Graham and Marvin, 2001).

In contrast, the relationship of ViDA with its surrounding environment is mediated by the residents' reactions to the context of violence in Monterrey. Fragmentation here happens from household to household as individual families try to securitize their home. Ideas of sustainability and green communities vanish in the face of rampant social violence (which has increased dramatically in Monterrey during the last decade associated with the growing interest of the drug cartels in the Monterrey 'plaza'). Violence encroaches in the daily life of residents: 'there is a lot of gang violence … you can see it here, in this development … not inside, inside there is not a lot but around … but I am here in the limiting area and I can see … they fight, they break the house … so we built a bit in the entrance, like a defense' (interview ViDA R4). Residents feel safe regarding the neighbours in the development, but still feel pressure to securitize their individual houses. This is done by building fences and annexes around the house, which protect the main living areas. Rather than using the collective playground, children play in front gardens now fenced up for each individual house. Also, as houses have become integrated in the neighbourhood lives and economies, they have been modified to provide security enclosures and by adding self-assembled structures to enable household-based commercial activities which are common in this type of settlement. The original designs — suited to the specific climatic conditions of the settlement — have also been modified by the integration of the development in the economic landscape.

Resource and land securitization, however, occurs only through a set of reconnections. In Bangalore, BCIL claims that T-Zed is a pioneering development which does not depend on municipal water supply thanks to an innovative design focused on rainwater harvesting. However, an unplanned long drought resulted in the system not being able to recharge itself as planned. Thus, BCIL built two bore wells in T-Zed at high depth, in the manner that privileged classes are already doing in other parts of the city. This allows the development to remain independent from the municipal piped supply but encroaches on collective groundwater resources. The original concept of maintaining the compound exclusively with rainwater harvesting had to be rethought: 'We have local water … we had to go and dig a few wells and withdraw the water, okay, that's a need out there. But water, water falls on to this piece of land … We make sure water that is harvested back out there' (interview T-Zed D1). Water harvesting is still done to complement the main supply, but the supply of water relies heavily on the existence of bore wells extracting from the collective water table in Bangalore (to which poorer sectors of society have no access). Moreover, this creates further tensions with people living in the surrounding areas. One T-Zed resident explained that:

> every morning there is a long line outside the gate, of people who want to … collect water, take water from T-Zed for drinking! … they don't have any drinking water! … And … the Association [of residents] doesn't approve of it! … because they don't know how many … once you start, you really can't stop! But, I think the security allows a lot of people in because either they pay [or] they empathise with them (interview T-Zed R4).

Residents have anxieties about what they perceive as an illegitimate use of the water resources they have securitized. However, the fundamental issue about T-Zed appropriating collective water resources and restricting access to other citizens is not discussed. T-Zed is embedded within wider discourses of appropriation of resources by the higher classes, which point to BCIL's ambiguous role in promoting sustainable and equitable development within Bangalore.

In ViDA, securitization practices have generated unease among the architects who designed the project because they see them as a violation of the design principles that inspired the project. New built areas disturb the visual appearance of the development, towards more fragmentation, and, they argue, obstruct schemes of natural ventilation and orientation that are supposed to help the house 'breathe'. This is a reflection of the different priorities of architects and residents. For example, in describing the ventilation windows, a resident said, with irony in his tone, that 'those windows are, most of all, to make dust because there is a lot of wind in this area' (interview ViDA R1). While upkeep practices have helped local residents to normalize the housing development within their set of expectations, they challenge the aspirations of those who see themselves as experiment makers and thus contest the need to showcase the experiment as a means to catalyze urban transitions towards low-carbon communities.

Finally, maintenance, as a resignification of the experiment, is a process that connects experiments into wider circulations of political economy and ecology. In Bangalore, T-Zed has served as a model to justify the adaptation of existing models of development to the moral requirements of the environmentally conscious high classes. The model is successful because it has been replicated by its promoters (BCIL) and because it has helped in the development of a low-carbon industry of consultants, suppliers and designers, which is behind the current policy and commercial push towards solar water collectors and rainwater harvesting in Bangalore.

The case of ViDA is particularly salient in this respect and deserves closer examination. While a field visit to ViDA suggests that the project has been almost forgotten on the ground, the project has become a central part of the national climate change discourse of Mexico, together with a handful of flagship social-housing projects. Policymakers have represented ViDA as a 'best practice' example for Infonavit's nationwide program 'Hipoteca Verde' (Green Mortgage). The Green Mortgage was a top-down response to the demand from Mexican leaders to do something to be shown for Mexico's international commitments to low-carbon policies. It is a provision for an additional amount to supplement existing mortgage packages in Infonavit dedicated to the incorporation of specific 'eco-technologies' (energy- and water-efficient appliances and ceiling insulation) sanctioned by official organisms and guidelines in standard social-housing templates. In 2011, 60% of Infonavit 480,000 mortgages (with an investment of 44,800 million Mexican pesos) will have a Green Mortgage. Since 2007, when the program started, Infonavit has given 380,000 Green Mortgages and they expect to make the program mandatory within the next few years.

Originally ViDA was conceived as an experimental project financed with conventional Infonavit mortgages and additional support from CONAVI and the Instituto de la Vivienda (to support low-carbon innovation in the design without disturbing existing templates). Infonavit launched the Green Mortgage as ViDA was underway and ViDA promoters saw this new policy as an opportunity to promote their design while incorporating energy- and water-efficient appliances, including low-energy bulbs, a small solar panel for the street lights and water-saving appliances for taps and showers. The appliance requirements of the Green Mortgage were then incorporated into the original design, and prospective owners were offered the Green Mortgage deal. ViDA was, in 2007, one of the first developments in Mexico to obtain a Green Mortgage but, in the overall life of the project, this was a last minute add on, on top of the bioclimatic design that inspired the project. Thanks to a series of expert evaluations and a prize (from the international cement corporation CEMEX), the project ViDA provided an example of how to incorporate low-carbon features without fundamentally changing the work patterns of developers or the relationships between residents, Infonavit and the developers. Moreover, the project ViDA demonstrated how additional capital could be redirected towards the creation of what were described as more sustainable forms of development. Now, ViDA's concerns of adapting developments to the local climate and economic conditions and its focus on how technologies benefit residents has been changed into 'a checklist of minimum technological requirements', as described by a local provider of energy-saving appliances (interview ViDA C4), that the house has to meet to qualify for the increase in the costs allowed by the Green Mortgage.

In that sense, the development has been appropriated at the national level as an example of how 'green building' can be adapted to the existing construction industry (rather than how the construction industry can be adapted to meet new low-carbon global demands). ViDA, together with other projects which also started relatively anonymously and have now become exemplars of green building in Mexico, has also generated new thinking about the potential for poor people to take responsibility for their — yet to occur — carbon emissions in an attempt to bring carbon control mechanisms beyond the realm of middle-class concerns. While in Bangalore T-Zed provided a space to experiment which has allowed the horizontal transmission of knowledge among consultants and service providers, sparkling similar projects at the same scale, in Monterrey, a relatively small and obscure example is at the centre of the reconfiguration of housing and infrastructure systems nationally, although the extent to which this is sustainable has to be discussed in the light of social and environmental impacts which the new houses have beyond their locale. Maintaining here has been directed towards reproducing the experimental character of the initiative and creating changes within existing urban regimes.

6.6 CONCLUSIONS

The case studies suggest that attempts to 'maintain' climate change experiments are central to their development through time. Following the analysis, maintaining can be understood as emerging from two interrelated processes. First, there is an ongoing process of repair management associated with the making of innovations and uncertain sociotechnical relationships which we have called upkeep. Second, there is a process of careful adjustment to existing urban conditions, physical (in terms of adapting to the specific elements of infrastructure and resources) and semiotic (in terms of developing suitable narratives of change and performance associated with the experiment), which we have called metabolic adjustment. Both processes contain an element of renegotiation of the mechanisms of space and nature production, in turn reconfiguring the realms of responsibility and authority in everyday maintenance, which determine the significance of the experiment for city- and nationwide low-carbon transitions.

The practice of repair is related to the ongoing work required to maintain the experiment 'as an experiment', in terms of sustaining its novel qualities (for example, repairing the different technological innovations in T-Zed) and in terms of normalizing it to ensure that it is an accepted way of life (whether it is redefining some components of the experiment, as in T-Zed, or assimilating them to everyday practices, as in ViDA). Maintenance is therefore needed to balance the novel and the normal within the experiment. While the discursive production of standards and expectations may accelerate this process, the pluricultural character of maintenance and its reliance on the power of things (Graham and Thrift, 2007) means that this is a difficult process that requires taming innovation and renegotiating original ideas, technologies, ecologies and relations in practice within the fabric of the city (cf. Kaika, 2005).

In terms of the process of adjustment, the experiment also requires a series of alternative disconnections, connections and reconfigurations to adjust it to the broader context within which it occurs and to establish its distinctiveness in that setting, whether it is dealing with water scarcity in T-Zed or managing contexts of violence and transforming policy discourses in ViDA. T-Zed and ViDA required specific processes to establish their long-term impact, in the case of T-Zed requiring the private developer to take an unorthodox role and in the case of ViDA hindered by the lack of mechanisms of engagement with local residents. Moreover, both cases signal a process of adjustment to specific neighbourhood conditions in terms of dealing with specific disputes and accommodating the needs of residents in T-Zed, or in terms of changing the designs to adapt to conditions of structural violence in ViDA. Finally, in both cases we observe a renegotiation of arrangements for the production of space and nature. In the case of T-Zed, notions of self-sufficiency are challenged by the reconnections with water resources and with neighbour residents (whether they are in other gated communities or in poor and informal settlements). In the case of ViDA, the role and history of the project have been renegotiated to adapt to the national demands of policymaking in sustainable building. Gaining visibility for infrastructure interventions is a key aspect of experiments. As with the process of normalization, in both cases, maintenance processes purposively seek to distance and distinguish the experiment as novel and to embed it within the existing sociotechnical frame. This balance between distance and embedding seems to be critical to the 'purchase' that the experiment gains in citywide networks of urban metabolism. In this way, the findings highlight the important role of infrastructure-based interventions in creating new reconfigurations of urban infrastructure and the maintenance of urban infrastructure regimes (Monstadt, 2009).

While the discourse of securitization is important in both examples, the social dynamics that shape them are very different. T-Zed exemplifies the attempts of elites to achieve urban ecological security (Hodson and Marvin, 2009), while ViDA demonstrates the importance of securing spaces for low-income groups in the context of structural violence. While in T-Zed security maintenance practices are key to ensure the continuity of the experiment and of the capacity of elites to influence the processes of urban reproduction, in ViDA security maintenance practices are directed towards coping with structural violence conditions. Because of the increased complexity and immediacy of maintenance practices (Graham and Thrift, 2007), who benefits from these practices will depend strongly on the context in which they are applied.

The experiments we have discussed here are anything but revolutionary. They do not seek to reconfigure the fundamental relations which structure society — their intention was never so. Nonetheless, we find that they serve as potentially powerful means to create new forms of governing everyday life and city circulations, simultaneously reinforcing and reinventing the landscape of governance. Critically, in giving the experiment its distinctive character and its ordinariness, maintenance is an essential component of how experiments travel. We find that such processes mean that the significance of the experiment does not end with its merging into the local landscape, as it can still be transformed into a symbol or exemplar with wider influence (whether among knowledge communities in Bangalore or at the national level in Mexico).

In conclusion, we find that the emergence of climate change experiments disrupts the sociotechnical system in such a manner that they require a process of 'maintaining' in which relationships around the experiments are redefined. They are embedded in the production of urban space and the transformation of nature as described in the urban political ecology literature. However, the process of 'maintaining' experiments is a precarious one which requires a continuous external input in terms of remaking the experiment (materially and in terms of redefining the narratives which maintain it); and may cause further reconfigurations beyond the experiment in terms of formal and informal institutions and changes in the configurations of responsibility and acceptability. However, this says nothing about the potential spaces for resistance and alternative appropriation of experiments, something which, we believe, is foregrounded in the processes whereby the experiments are lived as an integral part of everyday lives. How climate change experiments are made, maintained and lived is a question to which we intend to pay attention in our ongoing analysis of urban climate change experiments.

6.7 FOOTNOTES

1. In the context of climate change, urban change is understood as a 'transition' to a low-carbon society, that is, as a fundamental change in the way a service is delivered.
2. Throughout the article the interviews have been designated by a letter (developers D; consultants and knowledge producers C; residents R; public sector workers P; and NGOs N) and a number.

REFERENCES

1. Brand, S. (1994) How buildings learn: what happens after they're built. Penguin, New York.
2. Bulkeley, H. (2010) Cities and the governing of climate change. Annual Review of Environment and Resources 35.1, 229–253.
3. Bulkeley, H. and V. Castán Broto (2013) Government by experiment? Global cities and the governing of climate change. Transactions of the Institute of British Geographers 38.3, 361–375.
4. Castán Broto, V. and H. Bulkeley (2013) A survey of urban climate change experiments in 100 cities. Global Environmental Change 23.1, 92–102.
5. Cooke, J. and R. Lewis (2010) The nature of circulation: the urban political ecology of Chicago's Michigan Avenue Bridge, 1909–1930. Urban Geography 31.3, 348–368.

6. Desfor, G. and R. Keil (2004) Nature and the city: making environmental policy in Toronto and Los Angeles. University of Arizona Press, Tucson, AZ.
7. Gandy, M. (2004) Rethinking urban metabolism: water, space and the modern city. City 8.3, 363–379.
8. Ghosh, A. (2005) Public–private or a private public? Promised partnership of the Bangalore Agenda Task Force. Economic and Political Weekly XL, 4921–4922.
9. Graham, S. and S. Marvin (2001) Splintering urbanism: networked infrastructures, technological mobilities and the urban condition. Routledge, London.
10. Graham, S. and N.J. Thrift (2007) Out of order — understanding repair and maintenance. Theory, Culture and Society 24.3, 1–25.
11. Harvey, D. (1996) Justice, nature and the geography of difference. Blackwell, Oxford.
12. Heynen, N., M. Kaika and E. Swyngedouw (2006) Urban political ecology: politicizing the production of urban natures. In N. Heynen , M. Kaika and E. Swyngedouw (eds.), In the nature of cities: urban political ecology and the politics of urban metabolism, Routledge, London.
13. Hodson, M. and S. Marvin (2009) Urban ecological security: a new urban paradigm? International Journal of Urban and Regional Research 33.1, 193–215.
14. Hodson, M. and S. Marvin (2011) Can cities shape socio-technical transitions and how would we know if they were? In H. Bulkeley , V. Castán Broto , M. Hodson and S. Marvin , Cities and low carbon transitions, Routledge, London.
15. Hommels, A. (2005) Studying obduracy in the city: toward a productive fusion between technology studies and urban studies. Science, Technology, & Human Values 30.3, 323–351.
16. Hommels, A. (2008) Unbuilding cities: obduracy in urban sociotechnical change. MIT Press, Cambridge, MA.
17. Kaika, M. (2005) City of Flows. Routledge, London.
18. Keil, R. (2005a) Progress report — urban political ecology. Urban Geography 26.7, 640–651.
19. Keil, R. (2005b) Social power and the urbanization of water: flows of power. Annals of the Association of American Geographers 95.4, 889–891.
20. McFarlane, C. (2010) The comparative city: knowledge, learning, urbanism. International Journal of Urban and Regional Research 34.4, 725–742.
21. Monstadt, J. (2009) Conceptualizing the political ecology of urban infrastructures: insights from technology and urban studies. Environment and Planning A 41.8, 1924–1942.
22. Nair, J. (2000) Singapore is not Bangalore's destiny. Economic and Political Weekly XXXV, 1512–1514.
23. Ranganathan, M., L. Kamath and V. Baindur (2009) Piped water supply to Greater Bangalore: putting the cart before the horse? Economic & Political Weekly XLIV, 53–62.
24. Robinson, J. (2006) Ordinary cities: between modernity and development. Routledge, London.
25. Star, S.L. (1999) The ethnography of infrastructure. American Behavioral Scientist 43.3, 377–391.

26. Strebel, I. (2011) The living building: towards a geography of maintenance work. Social & Cultural Geography 12.3, 243–262.
27. Swyngedouw, E. (2006) Metabolic urbanization: the making of cyborg cities. In N. Heynen , M. Kaika and E. Swyngedouw (eds.), In the nature of cities: urban political ecology and the politics of urban metabolism, Routledge, London.
28. Swyngedouw, E. and N.C. Heynen (2003) Urban political ecology, justice and the politics of scale. Antipode 35.5, 898–918.
29. Thrift, N. (2005) But malice aforethought: cities and the natural history of hatred. Transactions of the Institute of British Geographers 30.2, 133–150.
30. Ward, K. (2006) 'Policies in motion', urban management and state restructuring: the trans-local expansion of business improvement districts. International Journal of Urban and Regional Research 30.1, 54–75.

CHAPTER 7

Assessment of Soil Sealing Management Responses, Strategies, and Targets Toward Ecologically Sustainable Urban Land Use Management

MARTINA ARTMANN

7.1 INTRODUCTION

The ongoing urbanization is one of the main threats for sustaining ecosystems' capability to supply ecosystem services to humans (MA 2005). Cities are characterized by a high degree of impervious surfaces and by continuous built-up areas (e.g., Turok and Mykhnenko 2007). Hence, urban growth promotes the increase in land take and soil sealing. Land take is understood as the conversion of open areas into built-up areas and can also include non-sealed areas such as gardens. Soil sealing is defined as the permanent covering of soil by completely or partly impermeable artificial material (Prokop et al. 2011). Sealing by urban gray infrastructure, which includes all forms of pavements and buildings (according to Breuste 2011), has especially negative impacts on the potential provision of ecosystem services. Soil sealing influences regulating services by increasing water surface runoff (Haase and Nuissl 2007) and microclimate regulation by increasing temperatures (Henry and Dicks 1987). It reduces provisioning services such as food production since fertile agricultural areas in particular get lost (Burghardt 2006). Furthermore, due to loss and fragmentation of habitats for flora and fauna, soil sealing has negative impacts on supporting services and is threatening urban biodiversity (Montanarella 2007). Moreover, the supply of cultural services is under pressure, since recreational areas within urban core districts are threatened by (re-)densifications (Niemelä et al. 2010).

Despite shrinking of the European population, a constant increase in impervious surfaces within the European Union can be observed (Prokop et al. 2011). The fact that there is a need to stop further soil sealing has already affected policies at the European and national level (EC 2012; EEA 2012). However, between 1990 and 2006 an increase of 8.8 % in artificial surfaces could be observed and, in 2006, 2.3 % of the European territory was sealed (Prokop et al. 2011). In Germany, a target was formulated which recommends the decrease of daily land take to 30 ha day−1 in 2020. This target seems difficult to reach since in 2010 still 77 ha day−1 were being taken for transport and settlement areas (Statistisches Bundesamt 2012). Neither a target on sealing reduction nor a standardized sealing monitoring exists in Germany. However, estimates show that between 46 and 50 % of transport and settlement areas are sealed (Breitenfeld 2009). Today, 5 % of the German territory is covered by impervious surfaces (Prokop et al. 2011).

Since cities in particular are characterized by a high degree of impervious surfaces, it is crucial to steer urban soil sealing in an ecologically sustainable way to secure urban ecosystems' ability to sustain ecosystem services for their residents. Therefore, this paper investigates (1) which responses, strategies, and sub-targets can be regarded as being efficient toward ecologically sustainable management of urban soil sealing and (2) who has competences and should take responsibility to steer soil sealing? Germany was chosen as the study area as this is one of the most sealed countries within the EU (Prokop et al. 2011). Table 1 provides definitions of the main terms used and their relation to the research questions.

7.1.1 SCALES OF INVESTIGATION AND STUDY AREA

The research integrates three scales, taking into account steering competences and addressees at the macro-, meso-, and microscale. At the macroscale regions (Region of Western Saxony/Region of Munich), federal states (Saxony/Bavaria) and the federal government (Germany) were considered. At the mesoscale, shrinking and growing cities with over 100 000 inhabitants in Germany were analyzed as it is assumed that the challenges cities face are especially complex due to the larger scale. Fourty-seven percent of European cities have a population of over 100 000 inhabitants (EC 2011). European cities are facing economic changes such as deindustrialization (Turok and Mykhnenko 2007), which offer cities the opportunity to re-use urban industrial wastelands to reduce further sealing. Moreover, European cities are confronted with social individualization, which leads to an increase in living space per capita. This hinders a reduction of land take (Haase et al. 2013) and decreases urban green areas (Kabisch and Haase 2013), which are essential for ecosystem service provision (Bolund and Hunhammar 1999). Growing and shrinking cities were differentiated because they face various challenges in urban management. Two case study cities were selected under specific selection criteria. Leipzig was selected between 1998 and 2008 as the highest increase in settlement and transport areas in Germany was recorded in the city despite the shrinkage processes. Munich was chosen due to a high increase in recreational areas between 1998 and 2008 and a high increase in population at the same time.

Leipzig is situated in Saxony, East Germany and has a population of 520 838 in 2012 (www.statistik.leipzig.de). Because of losing in economic importance in the 1960s, Leipzig experienced a high population migration. Despite processes of shrinkage, suburbanization, and urban sprawl could be observed, reaching their peak in the late 1990s (Haase and Nuissl 2010). Leipzig today is an example where both processes of shrinkage in the urban periphery and re-urbanization, especially in the urban core areas, can be found (Haase and Nuissl 2007). Previous studies on soil sealing development between 1997 and 2003 showed that sealing efficiency decreased during sealing at the urban fringes by commercial and industrial sites and low density residential areas. In total, an increase in sealed surfaces of 2.84 % could be observed and in 2003, 27 % of the area was sealed (Artmann 2013a).

With a population of 1.4 million (2011), Munich is the third largest city in Germany. Munich is characterized by a high immigration pressure as the population increased by over 200 000 residents between 1990 and 2010. Further population growth of 100 000 residents is projected by 2020. Moreover, Munich can be characterized by an urban re-organization due to the privatization of the German Railway System and the closing of barracks that were used for new residential and recreational areas. Compared to Leipzig, this supported a low increase of sealing, 0.4 % between 1998 and 2011 and in 2011, 36 % of the area was sealed. However, no further wastelands are available now and new residential areas should be built by further densification which threatens the loss of ecosystem services, especially in the urban core areas where green areas are already under pressure due to their small sizes and low per capita supply (Artmann 2013a). At the microscale, the civic society (NGOs and residents) and practitioners of relevance for soil sealing management (investors and (landscape) architects) were considered.

7.2 MATERIALS AND METHODS

7.2.1 DEFINITION OF SET OF INSTRUMENTS AND ITS STEERING COMPETENCES AND ADDRESSEES

To analyze the efficiency of soil sealing management strategies and sub-targets toward an ecological urban sustainable development (research question 1), sets of instruments considering a holistic soil sealing management approach were first defined. Sets of instruments were defined since German policy assumes that the 30-ha target can only be achieved by a mix of instruments (Deutscher Bundestag 2004). A holistic soil sealing management approach includes quantitative, qualitative, and compensatory management of urban gray and urban green as well as the protection of soils as the basis of urban gray and green. These steering dimensions are defined as sub-targets in this paper (Table 1) and were derived from a spatial analysis of soil sealing development (Artmann 2013a). For

Table 1. The 20 most populous MSAs in 2008 ranked by their deviation from the scaling law.

Term	Definition
Efficiency	Criterion of assessment which describes to which degree a response is suitable to achieve an objective in a certain way. The definition of an ecologically efficient soil sealing management approach is provided in Table 2
Response	Specific instrument which aims to steer soil sealing (e.g., a specific law such as the building code). This paper assesses the efficiency of ecological sustainable responses
Strategy	Strategy is understood as the sum of responses addressing the same types of steering. Within this study legal-planning, informal-planning, economic-fiscal, co-operative and informational strategies are investigated. The efficiency assessment of strategies is based on the assessment of responses which are assigned to strategies (see Fig. 1)
Sub-targets	Sub-targets define what has to be steered spatially in the course of a holistic soil sealing management approach. These targets relate to steering urban green (open land such as forests and agricultural land, recreational areas), gray (built-up areas and artificial material) and soil (land and substrate). Urban green and gray can be steered quantitatively (reduction of new sealing and land take, protection of green areas), qualitatively (promotion of internal development and space efficient building forms, protection of green areas with high ecological performance). Moreover, existing sealed areas can be compensated by de-sealing or greening roofs (see also Fig. 1). The efficiency assessment of sub-targets is based on the assessment of responses which are assigned to the sub-targets
Actors	Actors of soil sealing management refer to administrative units and communities responsible for developing and implementing strategies in the course of a holistic soil sealing management. The responses selected are assigned to groups of actors of different management scales to prove who is responsible for soil sealing management and to which degree (see Fig. 1)

assessing how these sub-targets can be achieved, strategies were identified by reviewing planning documents and literature. For soil sealing, relevant strategies include legal-planning (including laws and informal planning), economic-fiscal (e.g., subsidies, taxes), co-operative (e.g., regional or sectoral co-operations), and informational strategies (e.g., spatial monitoring, awareness raising, improving know-how) (Artmann 2013a). Specific instruments (named as responses) of each strategy were selected and assigned to the sub-targets via criteria (see Fig. 1).

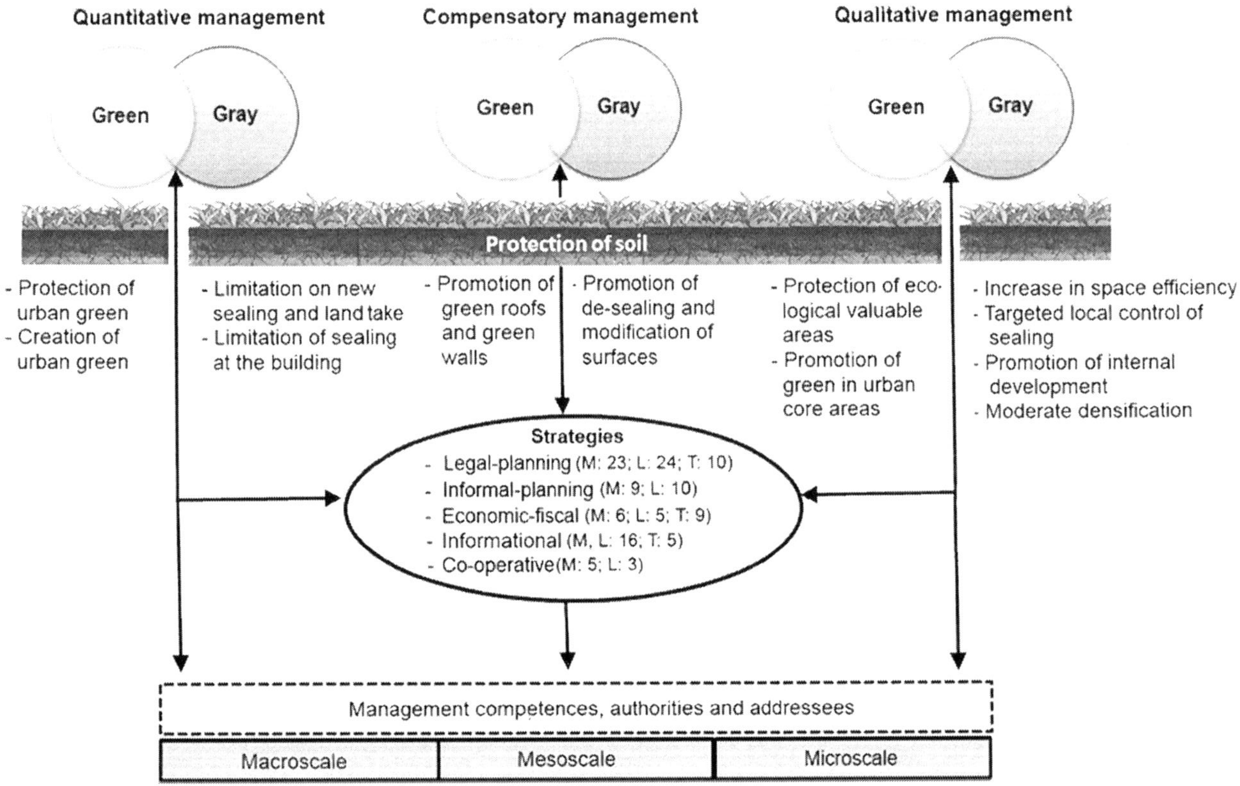

Figure 1. Framework for multi-scale analyses of soil sealing management instruments (M Munich, L Leipzig, T in theory discussed responses) (icon for soil by Osada 2011)

The selection of responses was done by reviewing laws as well as local-planning documents (zoning, landscape, sectoral, and informal plans), scientific literature and projects (such as REFINA, Research for the Reduction of Land Consumption and Sustainable Land Management), local initiatives and by conducting expert interviews including experts from the departments of planning, environmental reporting, environmental protection, urban redevelopment, and construction as well as NGOs, real estate agents and research. The responses selected should have relevance for steering soil sealing, land take and land use, urban green areas, and soil as part of a holistic soil sealing management. The focus of this paper is on responses in use. However, in further studies, theoretically discussed responses will be included. In total, 93 responses in practice and 24 theoretically discussed responses were identified and assigned to the sub-targets, whereas a response can be assigned to more sub-targets but only to one strategy. The number of responses selected per strategy is shown in Fig. 1. To analyze the main management authorities and addressees (research question 2), the selected responses were assigned to the macro- (state government, federal states, region), meso- (city level), and microscale (civic society, practitioners) by identifying who has the power to develop a response (authority) and who is responsible for implementing it (addressees) (see Fig. 1) (Artmann 2013b).

7.2.2 INDICATORS TO ASSESS STRATEGIES TOWARD AN ECOLOGICALLY SUSTAINABLE SOIL SEALING MANAGEMENT

The efficiency assessment of strategies and spatial sub-targets toward ecologically sustainable management was based on indicators. Indicators are useful as they support policy and decision makers by providing comprehensible and quick information on consequences of steering actions on the environment (Pulles and van Harmelen 2004). The indicators were derived by developing hypotheses of an ecologically sustainable development based on structured expert interviews in Leipzig and Munich, literature review and analyses of impacts by soil sealing on ecosystem services provisioning (Artmann 2013c). The indicators should reflect impacts of sealing on the urban ecosystem and ecosystem service provisioning as well as framework conditions for ecologically sustainable management.

7.2.2.1 INDICATORS ON IMPACTS BY SOIL SEALING ON ECOSYSTEM SERVICE SUPPLY

Ecosystem service supply strongly depends on land use. Therefore, indicators that assess the supply of ecosystem services should be sensitive to land use change (Larondelle and Haase 2013). Following this, land use policy steering urban soil sealing in an ecological sustainable manner should be aware of impacts on ecosystem service provision by soil sealing. This target becomes even more crucial as cities face global climate hazards (Bulkeley 2013) which are intensified by soil sealing. Moreover, according to experts in Leipzig and Munich, the increasing importance of "soft" location factors, including sufficient supply of recreational areas, improve the consciousness of impacts by sealing (Artmann 2013c). Recreational areas should offer characteristics such as "wilderness" or a "rich variety of species" (Herzele and Wiedeman 2003) and can thus be managed like urban forests. In contrast, urban parks are more managed (Bolund and Hunhammar 1999) but also provide physical and psychological well-being for urban dwellers (Chiesura 2004). Besides public green spaces, private green areas such as gardens and allotments are crucial for supporting urban biodiversity and for experiencing urban wildlife (Goddard et al. 2010). Spatial analyses of impacts on soil sealing in Leipzig between 1997 and 2003 showed that, in particular, soils of high quality were used for transport and settlement areas as part of the suburbanization processes (Artmann 2013a). The loss of valuable soils by sealing is crucial as fertile soils affect vital processes and functions such as nutrient cycling processes, seed dispersal, or pollination, which yield ecosystem services (Boyd and Banzhaf 2007). According to an expert of the Saxon State Office for the Environment, Agriculture, and Geology and a scientific expert, improved protection of agricultural areas could be promoted by stressing the importance of agricultural land for nutrition. To secure ecosystem services, the obligatory integration of ecological aspects and reduction of further sealing into decision making is crucial (Artmann 2013c).

7.2.2.2 INDICATORS ON FRAMEWORK CONDITIONS FOR A SUSTAINABLE ECOLOGICAL DEVELOPMENT

Spatial analyses in Munich on drivers of urban soil sealing between 1998 and 2011 showed that the main drivers of sealing were transport areas, which increased especially at the urban fringes (Artmann 2013a). In general, urban sprawl increases the distances between working and living and therefore the need for roads, which leads to an increase in the use of cars, energy consumption, and traffic emissions (de Ridder et al. 2008). Therefore, reducing private motorized traffic can support a reduction in sealing and at a larger scale also in energy consumption and air pollution (Artmann 2013a). Soil sealing management should therefore also include a spatial strategic overview and consider impacts by urban land use changes on distant rural places, also termed urban land teleconnections (Seto et al. 2012). Besides the spatial scale, a temporal hypermetropia is vital as the definition of sustainability in the Brundtland Report emphasizes achieving present development in a way which ensures that future generations can also meet their own needs.

7.2.3 THE ASSESSMENT PROCESS

The assessment of soil sealing management responses, strategies, and spatial targets toward an ecologically sustainable urban sealing management approach was based on a multi-attribute decision method (MADM) using an analytical hierarchy process (AHP). The MADM allows a comparison between several alternatives by using a set of indicators and therefore supports decision making (Zanakis et al. 1998). Within an AHP, one form of MADM, alternatives are compared in pairs including decision makers' preferences (Saaty and Vargas 2012). The assessment process included three steps: (1) assessment of importance of indicators, (2) content analysis of responses, and (3) evaluation of analyses results. More information on the method developed for Response-Efficiency-Assessment (REA) can be found in Artmann (2013b).

The assessment of the importance of indicators (step 1) was done by involving decision makers of the mesoscale responsible for urban development and planning, brownfield management, urban green management and nature conservation, soil sealing monitoring, urban renewal, and urban policy. In an online survey, the decision makers were asked to evaluate the importance of the indicators on a Likert Scale between 1 and 9, where 1 stood for not important and 9 for very important (see e.g., Mendoza and Prabhu 2000). The weighting factor W_I represents the mean value of the assessment (Table 2). The evaluation of the responses was carried out via a deductive content analysis (step 2), whereas the indicators served as a categorization matrix and were use to prove hypotheses developed before the analysis (Elo and Kyngäs 2008). Laws, planning documents, statements of initiatives, and co-operations were read carefully and data corresponding to the indicators excerpted. The excerpted passages were coded according to the indicators' assessment score (IS) for each sub-target (Table 2).

Afterward, the response efficiency (RE) was calculated for each response R of a strategy S separately for each sub-target ST (step 3): all indicator scores IS derived by the responses R within the spatial sub-targets ST (0)–(VI) were summed up and divided by the number N of responses R per strategies S reviewed. The sub-targets stand for (see also Fig. 1) (0) protecting soil; (I) quantitative steering urban gray; (II) quantitative steering urban green; (III) qualitative steering urban gray; (IV) qualitative steering urban green; (V) compensation measures for urban gray; and (VI) compensation measures for urban green. The quotient was multiplied by the weighting factor W_I (see Table 2):

$$RE_{ST-S} = \left(\frac{\sum_{I=1}^{12} IS_{R-RT}}{N_{R-S}} \times W_I\right) \quad (1)$$

The results are provided in % of the maximal reachable weighted score W_I per strategy S. For analyzing the most efficient strategy and spatial sub-targets toward an ecological sustainable soil sealing management, the percentage scores reached per strategy and sub-target were summed up, and the mean value for the strategies (5 strategies) and spatial targets (7 targets) was calculated.

Table 2. Indicators, assessment scores, and importance of indicators for assessing the efficiency of strategies toward ecologically sustainable soil sealing management (ES, Ecosystem service)

Indicator	Indicator assessment score IS (between 1 and 9)	Weighting factor W_I Munich (N = 13)	Weighting factor W_I Leipzig (N = 13)
Securing, improvement and development of habitats for flora and fauna	9: Protection of ES by securing green areas or soils/by reducing sealing is clearly stated as target interlinked with benefits derived by protection/reduction (e.g., reducing further sealing to protect habitats for flora and fauna and to improve contact to nature for residents) 7: Importance of ES/function is mentioned but not directly linked to targets such as reduction of further sealing/protection of green/soils (e.g., green areas are important for flora and fauna; sealed surface increase urban heating) 4: Protection/importance of aspects related to ES/functions are mentioned but they are not directly linked to benefits/harm by green areas/soils or sealing (e.g., measures for climate adaptation have to implemented, such measures could also integrate technical solutions) 1: ES not mentioned	6.85	7.00
Improving surface water run-off		7.08	8.13
Improving climate adaptation (decrease heat emission, increase carbon binding)		7.23	7.75
Improving private recreational areas (gardens, courtyards)		6.69	7.19
Improving public green areas (more managed areas such as parks)		6.69	6.44
Improving recreational areas (less managed, e.g., forests, landscape parks)		6.62	6.44
Protection of agricultural areas for food production		5.54	7.00
Protection of ecologically valuable fertile soils and their functions		7.00	7.44

Table 2. Continued.

Reducing motorized private transport	9: Demand for reduction of private motorized transport/development of public transport is mentioned related to the reduction of sealing/protection of green/ soils (e.g., the development of public transport is crucial to promote urban internal development) 7: Demand for reduction of motorized private transport/development of public transport is mentioned but not interlinked to targets for reducing sealing/protection of green/soils (e.g., further transport areas increase sealing) 4: Demand for reduction of impacts by motorized private transport/development of public transport are mentioned but not linked to reducing sealing/ protecting green areas (e.g., a decrease in motorized traffic reduces the air and noise pollution) 1: Demand for reduction of motorized private transport/development of public transport is not mentioned	6.54	7.50
Spatial strategic overview	9: supra-regional view; 7: regional view; 4: city view; 1: less than city view/ no spatial view	6.85	6.19
Temporal hypermetropia	9: >20 years; 7: 20–11 years; 5: 10–6 years; 3: 5–1 year; 1: no temporal course mentioned; 9: Integration ecological aspects before project implementation; 5: Integration ecological aspects during project implementation; 1: Integration ecological aspects after project implementation	6.46	7.31
Priority setting: Obligation for considering ecological aspects/reducing sealing or possibility of consideration	9: Reduction of further sealing/integration of ecological aspects is obligatory; 5: Reduction of further sealing/integration of ecological aspects is demanded but not binding as part of a weighing-up process with other aspects; 1: Ecological aspects are not mentioned at all	6.62	7.50

7.3 RESULTS

Figure 2 summarizes the average efficiency of strategies and spatial sub-targets toward an ecologically sustainable management as part of a holistic soil sealing management approach. In Munich and Leipzig, most of the responses analyzed focus on quantitative protection of urban green and qualitative steering of urban gray promoting infill development. Ecological arguments for infill development are a reduction of fragmentation of habitats, reduction of traffic, and protection of agricultural areas at the urban fringes. The protection of soil and the creation of green roofs are less often included in the reviewed responses. In Leipzig, de-sealing and dismantling measures are mentioned more often than in Munich especially in legal-planning and informal-planning documents to reduce transport areas, to adapt to climate change, and to improve recreational areas by demolishing buildings in highly densely built-up areas.

In Leipzig, as well as in Munich, steering by legal and informal planning seems to be the most efficient strategy followed by co-operative and informational strategies. The focus of legal and informal planning is especially on the promotion of urban infill development in course of a qualitative management of urban gray (Fig. 3). This sub-target is supported by the German Building Code (Baugesetzbuch) that allows a faster and more flexible realization of infill development waiving an environmental impact assessment. The aim is to protect natural areas and their fragmentation at the urban fringes. The quantitative reduction of further sealing is also demanded by the soil protection clause (Bodenschutzklausel), integrated in the German Building Code (Baugesetzbuch). Moreover, the open space plan of the City of Munich, the landscape plan of the City of Leipzig or the Regional Development plan of Bavaria and Saxony demand the reduction of further sealing, especially to improve microclimate regulation.

Economic-fiscal steering seems to be less efficient. This becomes especially obvious when looking at the economic-fiscal steering of soils for which no response could be identified (see Fig. 3). However, fiscal steering especially supports the promotion of inner development and the re-use of brownfields with the support of subsidies as well as the supply and quality of recreational areas. On the mesoscale, the city of Munich provides

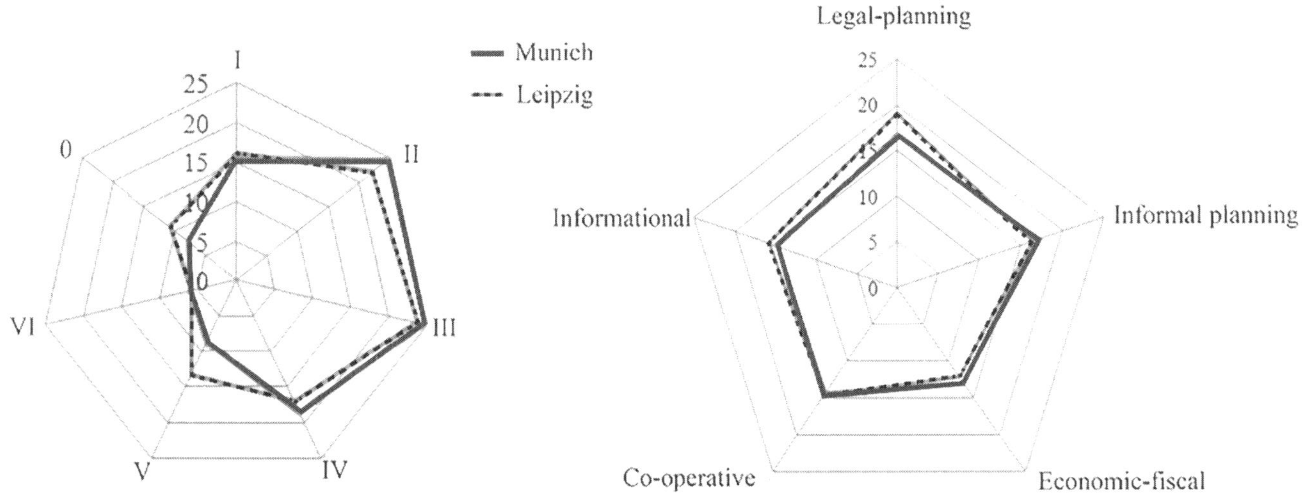

Figure 2. Spidergrams comparing the efficiency of spatial targets (left) and strategies (right) for ecologically sustainable soil sealing management (in %) in Leipzig and Munich. Sub-targets for steering soil sealing: 0 protecting soil; I quantitative steering urban gray; II quantitative steering urban green; III qualitative steering urban gray; IV qualitative steering urban green; V compensation measures for urban gray; VI compensation measures for urban green

financial support to practitioners and residents for de-sealing and greening roofs to improve the infiltration of surface water runoff and microclimate regulation as well as to improve living quality in the highly sealed city of Munich. Within a co-operative sealing management strategy, the focus in Leipzig is on the quantitative and qualitative steering of urban green which is especially supported by the regional co-operation Green Ring Leipzig (Grüner Ring Leipzig) where green areas are to be protected, established, and interlinked to protect agricultural areas and their fertile soils for food production, the development of parks for recreation or forests for improvement of biodiversity. Soft strategies such as co-operative and informational responses especially support greening roofs to protect ecosystem services. This results especially from the transfer of know-how about ecologic advantages of green roofs through brochures. As part of a participatory survey on the living quality in Munich residents demand a reduction of traffic, also in connection with the de-sealing of streets and the creation of green areas for recreation. The reduction of motorized traffic is also one sustainability target in Leipzig but focuses more on consequences by traffic on air pollution rather than on the space taken by cars. Moreover, within the sustainability targets of Leipzig, a reduction of further sealing is demanded. However, in Leipzig, no sealing monitoring exists as an informational strategy which could prove the target achievement. In Munich, monitoring of sealing exists but no quantitative targets exist corresponding to the monitoring.

The analyses of management competences and addressees showed that most of the reviewed responses are developed at the mesoscale, and therefore the city level has the highest competence to take appropriate steps to manage urban soil sealing (Fig. 4). The competence especially includes the quantitative (31 responses) and qualitative steering (30 responses) of urban gray. The qualitative steering of urban gray is also the most often addressed steering target by the state government (16 responses) and federal states (20 responses). However, compensation strategies, for instance, for urban green (state government: 5 responses, federal states: 4 responses) are more rarely developed at the macroscale but more often at the city level (19 responses). At the same time, the cities' policy and planning departments have to take responsibility to set the sub-targets through informal plans, monitoring systems, goals, co-operations between sectors as well as by putting laws stated by the state government into practice and acting as a role model to

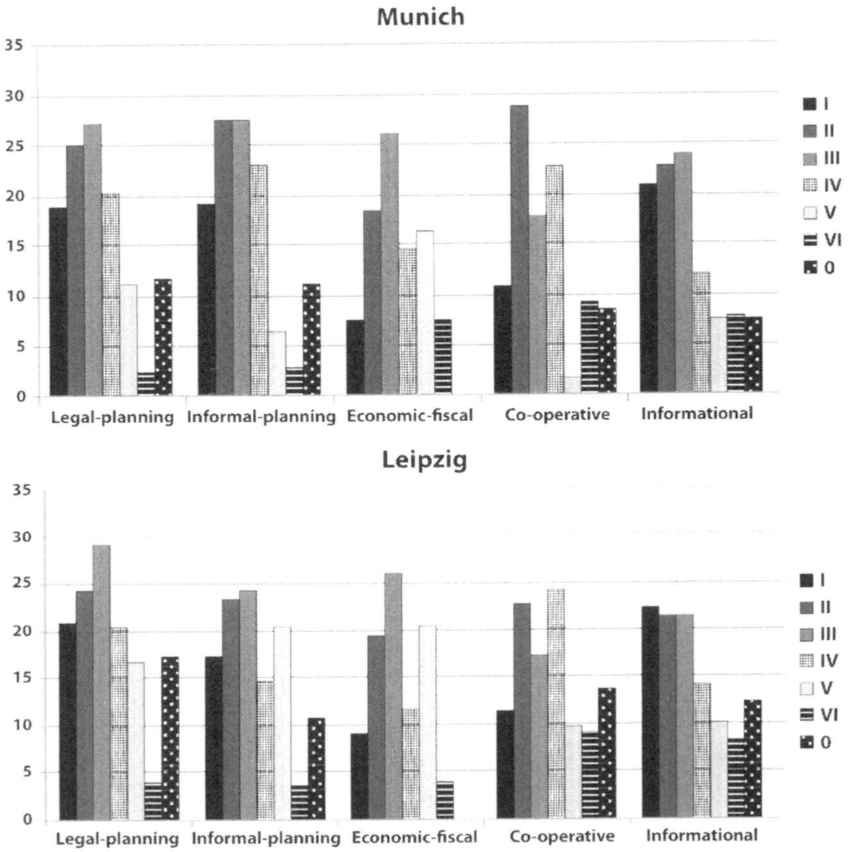

Figure 3. Efficiency of strategies to steer sub-targets of soil sealing management in Munich and Leipzig (in %). Sub-targets for steering soil sealing: 0 steering soil; I quantitative steering urban gray; II quantitative steering urban green; III qualitative steering urban gray; IV qualitative steering urban green; V compensation measures for urban gray; VI compensation measures for urban green

stop further loss of ecosystem services. Cities are especially addressed with respect to the implementation of strategies for qualitative steering of urban gray (28 responses) and quantitative steering of urban green (21 responses).

According to the review of responses, the microscale has developed responses less often as part of sealing management but is more important to put responses into practice. Therefore, the success of strategies and responses

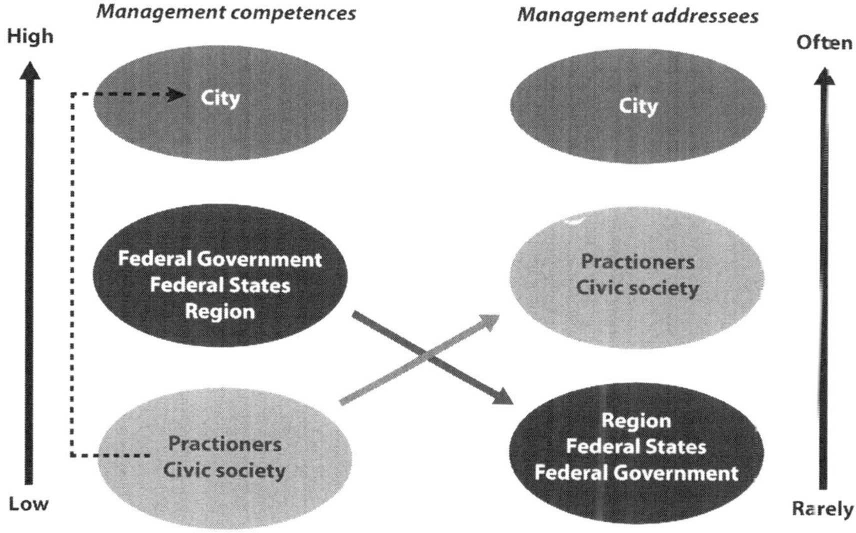

Figure 4. Management competences and addressees at the macro-, meso- and mircoscale based on the response analyses. The light diagonal arrow shows that actors at the microscale have less competence and are more often addressed for reducing soil sealing but influence the mesoscale (see dotted arrow). The dark diagonal arrow indicates that the macroscale has more competence but is less often addressed to implement any soil sealing measures

set by the city also depends on their steering potential on the microscale. For instance, subsidies for greening walls or taxes for waste water removal have to be high enough that residents implement such compensation measures. Informational strategies like brochures or consulting addressing the microscale could help to show up ecological and financial advantages of such actions.

7.4 DISCUSSION

7.4.1 EFFICIENCY OF SOIL SEALING MANAGEMENT TO SECURE ECOSYSTEM SERVICES

The evaluation of the responses showed that almost all strategies integrate ecological aspects as part of sealing management, which is also crucial in the course of global environmental change (Grimm et al. 2008). The importance of climate change within sealing management is also shown by the experts' highly ranked importance of indicators on improving surface water run-off and climate adaptation (Table 2) as well as the ranking of the most important indicators of the content analysis (Table 3). The adaptation to climate change could, according to the reviewed responses, especially be achieved by a quantitative steering of urban green. The positive effect of urban green on the microclimate has been well investigated in a range of studies (e.g., Gill et al. 2007; Jo and McPherson 2001). In contrast, the study showed that ecosystem services provided by fertile soils are less integrated in the responses reviewed. This might also occur due to the lack of scientific studies on soil and its provision of ecosystem services (see review within this special issue, Haase et al. 2014).

Moreover, the results demonstrated that the protection of soil by economic-fiscal strategies is missing in the case study cities in Germany. In other European countries, for instance, in Bulgaria or Poland, sealing of agricultural land is linked to a fee, the size of which depends on the quality of soil converted (EC 2012). However, although legal- and informal-planning strategies seem to be the most efficient, it cannot be confirmed within this study that a mix of economic-fiscal and land use planning instruments seems to be effective in reducing land consumption (Nuissl and Schroeter-Schlaack 2009), at least for steering soil sealing in an ecological manner.

The difference between the case study cities regarding the efficiency of steering was carried out by comparing the efficiency of spatial targets between Leipzig and Munich. The evaluation indicates the higher importance of de-sealing measures in Leipzig than in Munich due to high degrees of vacancy and brownfields that have arisen in the periods of shrinkage. In general, de-sealing is a chance to develop urban green areas, especially for Eastern European cities (Kabisch and Haase 2013). By investigating the development of urban green in 202 European cities, Kabisch and Haase (2013) further showed that an increase in living space per capita and in the number of smaller households hampers the reduction of further land take. Therefore, also practitioners and civic society need to be efficiently addressed to steer urban soil sealing; this also has been proven in this paper.

Table 3. Ranking of the three most important indicators per strategy for protecting ecosystem services in course of soil sealing management

Rank	Legal-planning		Informal-planning		Economic-fiscal		Informational		Co-operative	
	M	L	M	L	M	L	M	L	M	L
(1)	Climate	Climate	Climate	Climate	Water	Water	Climate	Climate	Climate	Less managed
(2)	Water	Water	Water	Less managed	Climate	Habitat	More managed	Water	Habitat	More managed
(3)	Habitat	Habitat	Less managed	More managed	More managed	Climate	Less managed	Less managed	Less managed	Habitat

M Munich, L Leipzig, climate improving climate adaptation, water improving surface water run-off, habitat securing of habitats for flora and fauna, more managed improving public green areas (more managed), less managed improving recreational areas (less managed)

7.4.2 FRAMEWORK CONDITIONS OF ECOLOGICALLY SUSTAINABLE SOIL SEALING MANAGEMENT

Indicators reflecting framework conditions for an ecologically sustainable soil sealing management approach showed that a temporal hypermetropia mainly achieves high scores for the indicators. Moreover, the majority of the responses reviewed considered at least ecological aspects but mostly only within a city view (Table 4).

Although there is a call that urban sustainability needs to consider planetary stewardship (Seitzinger et al. 2012), the review of the responses confirmed that urban policies neglect that urban development depends on the hinterland and its ecological and economical services (Rees 1992). For instance, experts in the boomtown of Munich evaluated that the protection of agricultural land for food production is less important than supporting regulating and cultural ecosystem services (Table 2). Also the review of strategies in both case study cities showed that the protection of this service is rarely implemented. However, the loss of agricultural land by land consumption means that food has to be transported into the cities, which might promote sealing for roads and increase air pollution by traffic, which then has global impacts. Moreover, the high space demand for motorized traffic is neglected by the responses reviewed. This shows the need to improve the know-how by scientists and decision makers about complex impacts caused by urban soil sealing and loss of open areas through urban land teleconnections (Seto et al. 2012).

7.4.3 LIMITS AND STRENGTHS OF THE STUDY

The evaluation of soil sealing management strategies and its efficiency toward an ecologically sustainable management approach was carried out by coding the results of the qualitative content analysis through numeric classes. This meant that weaknesses of a quantitative analysis, such as not seeing behind the scene of words and their meanings in a greater context (Selltiz et al. 1959), could be reduced. As the coding of the results is very context-sensitive, the process requires careful reading of the materials.

Table 4. Ranking of indicators per strategy regarding framework conditions for ecologically sustainable steering of soil sealing

Rank	Legal-planning		Informal-planning		Economic-fiscal		Informational		Co-operative	
	M	L	M	L	M	L	M	L	M	L
(1)	Temp.	Temp.	Temp.	Temp.	Temp.	Temp.	Spat.	Spat.	Oblig.	Oblig.
(2)	Oblig.	Oblig.	Spat.	Oblig.	Oblig.	Spat.	Temp.	Temp.	Temp.	Spat.
(3)	Spat.	Spat.	Oblig.	Spat.	Spat.	Red.	Oblig.	Oblig.	Spat.	Red.

M Munich, L Leipzig, temp. temporal hypermetropia, oblig. obligation for considering ecological aspects/reduction of sealing, spat. spatial strategic overview, red. reducing motorized private transport

Therefore, the content analysis was repeated twice. By using indicators and coding, the results to compare strategies and spatial sub-targets with each other, the approach presented in this paper complements the qualitative and less systematic analysis of best practice examples for soil sealing management by the European Commission (EC 2012). However, a pairwise ranking of each response by experts, like in a traditional AHP, was not done due to the high amount of responses studied; but at least experts were involved in developing the indicators and assessing their importance.

Limits in this study occurred through the selection of responses as it cannot be guaranteed that all responses handling soil sealing in the case study cities have been included. However, as experts were consulted in the identification of important responses and several studies on urban soil sealing and land take were revised, it can be assumed that especially legal-planning, informal-planning, and economic-fiscal strategies are complete as these provide the basis for urban development. However, informational and co-operative strategies might be incomplete as a range of small local initiatives could exist, such as civic greening communities (Bendt et al. 2013), which were not integrated separately into the study. However, the most important co-operations like the green ring Leipzig were elaborated. Further research is necessary to investigate to which degree such local greening initiatives support, for instance, the protection of urban green areas and the stewardship of ecosystem services, as undertaken by Bendt et al. (2013).

7.4.4 IMPORTANCE OF FINDINGS FOR RESEARCH ON ECOSYSTEM SERVICES

Although responses for reducing sealing and protecting green for the provisioning of ecosystem service were identified, none of these mentioned the term "ecosystem services." This has been shown in a study in Finland where most of urban land use planning actors were not familiar with the concept of ecosystem services although aspects of it were included in land use planning (Niemelä et al. 2010). Therefore, a closer co-operation between science and practice seems crucial to promote the concept of ecosystem services. This might improve a comprehensive understanding of municipalities and their inhabitants regarding the ecosystem and the benefits they derive from it for their well-being (Daily et al. 2009; Niemelä et al. 2010).

Findings of the study also showed that the qualitative steering of soil sealing is primarily understood as the promotion of infill development and densification neglecting that a sufficient supply of ecosystem services is also crucial for living quality in urban core areas as urban ecosystem services should be provided where they are consumed (Bolund and Hunhammar 1999). In this regard, the ecosystem service approach can provide decision support for policies to identify which green areas should be protected from further sealing and where sealing would be acceptable by assessing the supply and demand of ecosystem services. However, further research is necessary to provide standardized methods and indicators for planning and policy assessment, which are practical, applicable, comprehensive, credible, sensitive to changes in land management as well as temporarily and spatially explicit (van Oudenhoven et al. 2012).

7.5 CONCLUSION

The paper introduced a new analytical approach to assess and compare strategies and spatial sub-targets to secure ecosystem services using the example of soil sealing management. It contributes to a clearer understanding about which ecosystem services are considered by planning and policy to be threatened through soil sealing and land consumption and which have to be secured by protecting urban green and soils. It could be shown that challenges as a result of climate change such as improvement of microclimate regulation and reduction of floods are the most important arguments to reduce further sealing and to protect urban green. However, the study responds to the increasing need to include the soil as an ecosystem service provider in further research as well as to detect complex connections between ecosystem service provision and land use change as part of urban land teleconnections. Nevertheless, this study showed that the basis for an ecological sustainable management of urban soil sealing steering is assured, especially by legal- and informal-planning strategies. However, since sealing is further increasing in Europe, it can be concluded that strategies lack efficient implementation. Therefore, further research should focus on assessing the steering potential of these responses (e.g., acceptance of responses, control of steering success) by integrating actors of the meso- and microscale (as the main steering addressees) into the assessment process.

REFERENCES

1. Artmann, M. 2013a. Spatial dimensions of soil sealing management in growing and shrinking cities—A systemic multi-scale analysis in Germany. Erdkunde 67: 249–264.
2. Artmann, M. 2013b. Response-efficiency-assessment—A conceptual framework for rating policy's efficiency to meet sustainable development on the example of soil sealing management. Journal of Environmental Assessment Policy and Management 15: 135002.
3. Artmann, M. 2013c. Driving forces of urban soil sealing and constraints of its management—The cases Leipzig and Munich (Germany). Journal of Settlements and Spatial Planning 4: 143–152.
4. Bendt, P., S. Barthel, and J. Colding. 2013. Civic greening and environmental learning in public-access community gardens in Berlin. Landscape and Urban Planning 101: 18–30.
5. Bolund, P., and S. Hunhammar. 1999. Ecosystem services in urban areas. Ecological Economics 29: 293–301.
6. Boyd, J., and S. Banzhaf. 2007. What are ecosystem services? The need for standardized environmental accounting units. Ecological Economics 63: 616–626.
7. Breitenfeld, J. 2009. Methodology for calculating soil sealing and space productivity and evaluation of federal states environmental economic accounts (UGRdL) (in German). IÖR-Workshop Spatial Monitoring. Retrieved January 21, 2009, from http://www.ioermonitor.de/fileadmin/Dokumente/Symposium_2009_Expertenworkshop/breitenfeld.pdf.
8. Breuste, J.H. 2011. Ecology in cities: man-made physical conditions—Summary. In Urban ecology. Patterns, processes, and applications, ed. J. Niemelä, 71–72. New York: Oxford University Press.
9. Bulkeley, H. 2013. Cities and climate change. Oxon: Routledge.
10. Burghardt, W. 2006. Soil sealing and soil properties related to sealing. In Function of soils for human societies and the environment, ed. E. Frossard, W.E.H. Blum, and B.P. Warkentin, Special Publications 266, 117–124. London: The Geological Society.
11. Chiesura, A. 2004. The role of urban parks for the sustainable city. Landscape and Urban Planning 68: 129–138.

12. Daily, G.C., S. Polasky, J. Goldstein, P.M. Kareiva, H.A. Mooney, L. Pejchar, T.H. Ricketts, J. Salzman, et al. 2009. Ecosystem services in decision making: Time to deliver. Frontiers in Ecology and the Environment 7: 21–28.
13. De Ridder, K., F. Lefebre, S. Adriaensen, U. Arnold, W. Beckroege, C. Bronner, O. Damsgaard, I. Dostal, et al. 2008. Simulating the impact of urban sprawl on air quality and population exposure in the German Ruhr area. Part II: Development and evaluation of an urban growth scenario. Atmospheric Environment 42: 7070–7077.
14. Deutscher Bundestag. 2004. Report by the Federal Government on the perspectives for Germany—National strategy for a sustainable development. Printed matter 15/4100, Berlin: German Bundestag (in German).
15. EC (European Commission). 2012. Guidelines on best practice to limit, mitigate or compensate soil sealing. Commission staff working document, Brussels, Belgium, 65 pp.
16. EC (European Commission, Directorate General for Regional Policy). 2011. Cities of tomorrow—Challenges, visions, ways forward. Publ. Office of the European Office, Luxembourg, 116 pp.
17. EEA (European Environment Agency). 2012. The State of Soil in Europe. A contribution of the JRC to the EEA Environment State and Outlook Report—SOER 2010, European Environment Agency, Joint Research Centre, Luxembourg, 80 pp.
18. Elo, S., and H. Kyngäs. 2008. The qualitative content analysis process. Journal of Advanced Nursing 62: 107–115.
19. Gill, S.E., J.F. Handley, A.R. Ennos, and S. Pauleit. 2007. Adapting cities for climate change: The role of the green infrastructure. Built Environment 33: 115–133.
20. Goddard, M.A., A. Dougill, and T.G. Benton. 2010. Scaling up from gardens: Biodiversity conservation in urban environments. Trends in Ecology & Evolution 25: 90–98.
21. Grimm, N.B., S.H. Faeth, N.E. Golubiewski, C.L. Redman, J. Wu, X. Bai, and J.M. Briggs. 2008. Global change and the ecology of cities. Science 319: 756–760.
22. Haase, D., and H. Nuissl. 2007. Does urban sprawl drive changes in the water balance and policy? The case of Leipzig (Germany) 1870–2003. Landscape and Urban Planning 80: 1–13.
23. Haase, D., and H. Nuissl. 2010. The urban-to-rural gradient of land use change and impervious cover: A longterm trajectory for the city of Leipzig. Journal of Land Use Science 5: 123–141.
24. Haase, D., N. Kabisch, and A. Haase. 2013. Endless urban growth? On the mismatch of population, household and urban land area growth and its effects on the urban debate. PLoS ONE 8: e66531.
25. Haase, D., N. Larondelle, E. Andersson, M. Artmann, S. Borgström, J. Breuste, E. Gomez-Baggethun, Å. Gren, et al. 2014. A quantitative review of urban ecosystem service assessments: Concepts, models, and implementation. AMBIO 43. doi:10.1007/s13280-014-0504-0.
26. Henry, J.A., and S.E. Dicks. 1987. Association of urban temperatures with land use and surface materials. Landscape and Urban Planning 14: 21–29.

27. Herzele, A.V., and T. Wiedemann. 2003. A monitoring tool for the provision of accessible and attractive urban greenspaces. Landscape and Urban Planning 63: 109–126.
28. Jo, H.K., and E.G. McPherson. 2001. Indirect carbon reduction by residential vegetation and planting strategies in Chicago, USA. Journal of Environmental Management 6: 165–177.
29. Kabisch, N., and D. Haase. 2013. Green spaces of European cities revisited for 1990–2006. Landscape and Urban Planning 110: 113–122.
30. Larondelle, N., and D. Haase. 2013. Urban ecosystem services assessment along a rural–urban gradient: A cross-analysis of European cities. Ecological Indicators 29: 179–190.
31. MA (Millenium Ecosystem Assessment). 2005. Ecosystems and human well-being: Biodiversity synthesis. World resources Institute, Washington, DC, U.S., 100 pp.
32. Mendoza, G.A., and R. Prabhu. 2000. Development of a methodology for selecting criteria and indicators of sustainable forest management: A case study on participatory assessment. Environmental Management 26: 659–673.
33. Montanarella, L. 2007. Trends in Land Degradation in Europe. In Climate and land degradation, ed. M.V.K. Sivakumar, and N. Ndiang'ui, 83–104. Berlin: Springer.
34. Niemelä, J., R.S. Saarela, T. Söderman, L. Kopperoinen, V. Yli-Pelkonen, S. Väre, and D.J. Kotze. 2010. Using the ecosystem services approach for better planning and conservation of urban green spaces: A Finland case study. Biodiversity and Conservation 19: 3225–3243.
35. Nuissl, H., and C. Schroeter-Schlaack. 2009. On the economic approach to the containment of land consumption. Environmental Science & Policy 12: 270–280.
36. Osada, D. 2011. Impact of the pH-value on the soil life and nutrient incorporation (in German). Retrieved August 30, 2013, from www.ihr-einkauf.com/magazin/de/einfluss-des-ph-wertes-auf-bodenleben.
37. Prokop, G., H. Jobstmann, and A. Schönbauer. 2011. Report on best practices for limiting soil sealing and mitigating its effects. Study contracted by the European Commission, DG Environment, Technical Report-2011-50, Brussels, Belgium, 231 pp.
38. Pulles, T., and T. van Harmelen. 2004. General policy questions need specific policy effectiveness indicators: An example for emission reduction targets. Environmental Sciences 1: 351–367.
39. Rees, W.E. 1992. Ecological footprints and appropriated carrying capacity: What urban economics leaves out. Environment and Urbanization 4: 121–130.
40. Saaty, T.L., and L.G. Vargas. 2012. Models, methods, concepts and applications of the analytic hierarchy process. New York: Springer.
41. Seitzinger, S.P., U. Svedin, C. Crumley, W. Steffen, S.A. Abdullah, C. Alfsen, W.J. Broadgate, F.H.B. Biermann, et al. 2012. Planetary stewardship in an urbanising world: beyond city limits. AMBIO 41: 787–794.
42. Selltiz, C., J. Deutsch, and S.W. Cook. 1959. Research methods in social relations. New York: Holt, Rinehart and Winston.
43. Seto, K.C., A. Reenberg, C.G. Boone, M. Fragkias, D. Haase, T. Langanke, P. Marcotullio, D.K. Munroe, et al. 2012. Urban land teleconnections and sustainability.

Proceedings of the National Academy of Sciences of the United States of America 109(20): 7687–7692.

44. Statistisches Bundesamt. 2012. Sustainable development in Germany. Indicator report 2012. Wiesbaden: Statistisches Bundesamt (in German).
45. Turok, I., and V. Mykhnenko. 2007. The trajectories of European cities, 1960–2005. Cities 24: P165–P182.
46. van Oudenhoven, A.P.E., K. Petz, R. Alkemade, L. Hein, and R.S. de Groot. 2012. Framework for systematic indicator selection to assess effects of land management on ecosystem services. Ecological Indicators 21: 110–122.
47. Zanakis, S.H., A. Solomon, N. Wishart, and S. Dublish. 1998. Multi-attribute decision making: A simulation comparison of select methods. European Journal of Operational Research 107: 507–529.

CHAPTER 8

Developing a Sustainability Assessment Model: The Sustainable Infrastructure, Land-Use, Environment and Transport Model

TAN YIGITCANLAR AND FATIH DUR

8.1 INTRODUCTION

In recent years climate change and other rising environmental concerns and problems have put sustainable urban development on the top of the agenda in almost every city across the world [1,2]. The increased and urgent environmental agenda has engendered the need for employing sustainability assessment frameworks as key mechanisms for measuring the impacts of development on the environment, and as key policy instruments for supporting the transition to a sustainable urban development path [3,4]. Particularly during the last decade, sustainability assessment via indicators and indexing methods has gained recognition. This is mainly because of the visualisation of phenomena and the highlighting of trends based on reliable variables being highly considered as logical approaches in determining comparative sustainability levels [5]. Among the experts, there is a common agreement on using sustainability indicators for assessment, provided that they are selected and applied carefully and appropriately [6]. Indicators help in the performance assessment of the development, and of the overall effectiveness of partnerships to improve economic, social and environmental well-being of urban settings. Beyond the assessment purpose, sustainability indicators are also crucial in developing awareness of urban and environmental problems, and in advocating the need for achieving sustainable urban development [7].

Following the wide acceptance of the sustainable urban development notion, finding an accurate way to assess and measure comparative sustainability levels of existing and future developments has become an important issue [8,9], and, there have been various studies which have proposed different methods for sustainability assessment [10-13]. A thorough review of some of these assessment tools are presented by Karol and Brunner, particularly scrutinising six key neighbourhood scale sustainability assessment tools—the Cascadia Scorecard, the LEED (Leadership in Energy and Environmental Design) for Neighbourhood Development Rating System, One Planet Living (OPL), the South East England Development Agency (SEEDA) Checklist, SPeAR® (Sustainable Project Appraisal Routine), and VicUrban Master Planned Community Assessment Tool [14].

Although there are various sustainability assessment methodologies, models and tools developed so far, only a few have an integral approach that takes into account all of the environmental, economic and social aspects. According to Singh et al. [5] "...in most cases the focus is on one of the three aspects. Although, it could be argued that they could serve supplementary to each other, sustainability is more than an aggregation of the important issues, it is also about their inter-linkages and the dynamics developed in a system. This point will be missing if tried to use them supplementary and it is one of the most difficult parts to capture and reflect in measurements..."

Hacking et al. [15] advocate that the confusion inherent in sustainability assessment methods might be avoided by gathering all these methods under a broad umbrella of "sustainability assessment appraisal" and forming a more precisely defined method based on sustainability indicators. The literature indicates limitations of the existing assessment models and sustainable urban development requests, which are rapidly increasing in sophistication, and this creates an urgent need for more effective assessment methods and tools [16,17].

In line with these sentiments, this paper reports a study that develops a comprehensive sustainability assessment model entitled "The Sustainable Infrastructure, Land-use, Environment and Transport Model (SILENT)," which aims to provide a more effective sustainability assessment by taking all of the major aspects affecting sustainability into account: environmental, economic and social. The SILENT Model is developed as an advanced geographic information system and indicator-based urban sustainability indexing model. The model considers the sustainability of land-use, environment, transport systems and infrastructure with a triple bottom line approach, and uses similar steps of the OECD Composite Indicators Methodology [18].

This paper is organised in four sections. Following this introduction, secondly, we highlight the nature and importance of indicator-based comparative urban sustainability assessment, which is increasingly recognised as a successful sustainability assessment method. Thirdly, we present the conceptual and methodological approach of the SILENT Model. In this section, we also introduce the four key components of the model: conceptual, indicator, indexing and policy-support bases of the model. Lastly, we outline potential policy implications and plans for further development of the SILENT Model.

8.2 INDICATOR-BASED COMPARATIVE URBAN SUSTAINABILITY ASSESSMENT

As Meadows [19] notes, indicators arise from values and, in turn, they create values; therefore, the biggest advantage of an indicator-based comparative urban sustainability assessment model is the quantifiability of the comparative sustainability levels. Another instrumental purpose of using sustainability indicators is that "...by visualising phenomena and highlighting trends, indicators simplify, quantify, analyse and communicate otherwise complex and complicated information..." [20]. Depending on these qualities, indicators have attracted a wide range of interest, and this has led to generation of a large number of relatively successful urban sustainability assessment practices. The main difficulty faced while using indicators is to find a common unit and method of measurement leading to comparison of performance of a setting or policy package. Over the last decade, there has been an increasing effort to structure an indicator system and monitoring process to accurately formulate an integrated urban sustainability monitoring and assessment strategy.

In such a strategy, even though the ecological footprint is not a composite indicator, because both composite indicators and the ecological footprint collapse all sustainability issues into a single number, the ecological footprint approach is considered useful in finding a common unit for measurement [21]. The ecological footprint documents the extent to which human economic activities stay within the regenerative capacity of the biosphere. It considers consumption or production perspectives related to the activities of cities, corporations and individuals, and their role in non-renewable resource depletion. This approach is popular because its standardised measurement—global hectares—can be employed when making urban and project based comparisons [22]. This generalised and comparable measurement is often used by public and private institutions, and this is seen as a positive development in urban policy analysis [23]. As emphasised by McManus and Halighton [23], the ecological footprint concept contributes to focusing on and minimising consumption patterns and changing global aspects of these patterns.

From a local perspective, sustainability indicators reflect large scale environmental and economic considerations, as well as local social issues relevant to urban sustainability. In general, catchments, habitats of endangered species and natural reserve areas form environmentally sensitive regions, and environmental sustainability considerations are highlighted at the local and regional scales. In terms of economic activities and urban communities, a divergent range of spatial units, such as; from metropolitan areas to small scale infill areas, are the main subjects of local level sustainability studies. In these studies, there is a growing concern to balance environmental, economic and social dimensions of sustainability [24]. The sustainability of local economy, residential and industrial consumption, recycling, energy security, renewable energy use, local pollution, preservation of ecologically sensitive areas, accessibility to urban services, demographic changes, immigration and integration of social groups, social and gender equity, urban poverty, quality of life, sense of community, public security, participation in local decision making process, education, literacy, and public health are among the key indicator categories that can be found in nearly all sustainability assessment endeavours at the local level.

The scope and contents of local indicators differ from one project to another. However, the primary intention of a sustainability assessment is to include the most prominent local indicators in the assessment model. An assessment model with a comprehensive coverage of key issues provides findings that, in an extensive and inclusive decision making platform, could support the development of policies and actions for a more sustainable urban future [25-27].

In order to measure the comparative level of sustainability accurately, sustainability indicators ought to be carefully selected. On the theoretical front, indicators should relate to sustainability and represent all necessary sustainability domains (i.e., economic, environmental, social, and also institutional). On the practical front, they should have appropriate parameters that would make assessment possible. Lautso et al. [28] define the key indicator selection issues as relevance, representativeness, policy and predictability. Furthermore, indicators need to be scientifically valid, responsive to the changes in the system, understandable, and flexible enough to encompass new knowledge and public perceptions [29]. In relation to the data availability and quality, indicators should be as few as possible; however, no essential indicator should be omitted for purposes of brevity. Even if these qualities are context-dependent and not exhaustive, as Hak et al. [30] state, "…features of a robust indicator include a simple and unified method, commonly agreed issues and targets of wide applicability, transparency in the process, and agreement between partners on the process…"

Most indicator-based approaches only highlight issues, and do not provide an answer to the question of why the level of sustainability differs from one place to another. In other words, in most of these approaches, the link between theory and practice has not been well established. Therefore, it is important that key indicators need to be supplemented by qualitative and quantitative information on impact and performance levels. In this regard, a good sustainability assessment practice example is The Bellagio

Principles developed by the International Institute of Sustainable Development [2]. These principles serve as guidelines for the assessment process, including the selection and design of indicators, their interpretation, and the communication of results, to provide a link between theory and practice.

8.3 THE SILENT MODEL

Unsustainable urban development occurs mostly as a result of the inherent interdependence between urban form, transportation and infrastructure, and their impacts on the environment. Therefore, for this research our operational definition of urban sustainability is "the long term viability of urban living that minimises the negative impacts of urban demography, land use, urban form and transport on the environment." As explained in the previous sections, sustainability assessment plays an important role in portraying unsustainable urban development as well as good sustainability practices. Considering the constrains and limitations of existing sustainability assessment methods, and the tools depicted and discussed in the literature, this paper reports a study that develops a local level comprehensive sustainability assessment model. "The Sustainable Infrastructure Land-use Environment and Transport Model (SILENT)" is an advanced geographic information system (GIS) and indicator-based urban sustainability indexing model. The spatial indexing nature of the model is particularly useful for the analysis and the visualisation of comparative sustainability levels of urban localities. As a spatial indexing endeavour, the specific aim of the model is to incorporate all related domains affecting urban sustainability (i.e., demography, land-use, environment, transport and infrastructure) into a practical assessment method that informs planning and decision making processes. The SILENT Model is developed by following four logical steps similar to the OECD's Composite Indicators Methodology [18].

Firstly, a relevant measurement method to evaluate performance of urban sustainability is selected. As the background provided in the earlier literature review sections suggests, an indicator-based sustainability approach is selected due to its conceptual consistency and practical simplicity. Additionally, the SILENT Model uses performance indicators selected from the current planning schemes that reflect local sustainability concerns (i.e., Gold Coast City Planning Scheme's sustainability indicators). Secondly, to gauge the comparative sustainability levels of the urban settings, a number of indicator categories, sets, individual indicators and parameters are employed (see Appendices 1). By using relevant indicators, the model analyses and pictures the comparative sustainability levels mainly based on the composite effects of urban form, transport and infrastructure interrelationship and their impacts on the environment. The third step is aggregating the values of each area by individual indicators to form a composite index. This step includes associating different indicators to form a composite index by assigning and considering each indicator's individual weightings. To do this, a number of statistical procedures are applied. Although at the first glance the SILENT Model looks like only a composite indexing system, actually it also benefits from multivariate analysis in forming the indicator base of the model. At this initial practice, reported in this paper, only factor analysis is employed as a multivariate analysis technique, however, along with factor analysis a stepwise regression method is also possible to be used in order to determine the best fitting sets of indicators to identify sustainability levels. Finally, the model outputs—the spatial sustainability composite index values—are prepared for use in the benchmarking and policy making processes. As well as revealing the existing comparative levels of urban sustainability, the model can also be used to estimate the sustainability outcomes of alternative development scenarios.

The main characteristic of the SILENT Model is that it uses a grid-based system and divides the study area into small grid cells (e.g., 100 × 100 m). The grid-based analysis is particularly popular in accessibility indexing studies due to its strengths in condensing the analysis into comparable same size analysis units—for example, LUPTAI [31]. The model assigns values of various attributes of urban settings into each grid cell by using an indicator-based assessment system. The completion of value assignment to each grid cell for each indicator forms a composite index in both tabular and dynamic visual forms (that are tables and GIS maps) to display the results of the sustainability assessment performance of the locality. The model is also equipped to be run for policy and scenario testing.

The nature of the iterative use of the SILENT Model provides a dynamic and process-dependent sustainability evaluation.

The structure of the SILENT Model is illustrated in Figure 1 below. There are four constituent parts of the model: conceptual base of the model; construction of the indicator base of the model; urban sustainability indexing base of the model; and the policy and decision support base of the model. These constituent parts of the model are explored in detail below.

The concept of sustainability and its spatial or urban structure dimension constitute the theoretical foundation of the SILENT Model. In terms of sustainable urban development and sustainable communities, urban form, mobility pattern and infrastructure provision are the primary issues connected to the environmental domain of sustainability. Urban policy considerations are generally delineated by land-use and transportation plans and infrastructure investments. Naturally, all endeavours related to urban development carry infrastructure and service considerations into the planning activities. Therefore, the question remaining is: how to define and measure the interrelated qualities of this construct to portray interventions designed to form more sustainable communities. Indicators and indices are frequently used means for generating sustainability policies and making comparison among different aspects of the sustainability performance. Even if they are widely used tools, the theory behind the indicator-based description of urban sustainability (with scientific reasoning) frames the structure of the research and has immense importance for the robustness and reliability of methods. Even though there is no unified method in the indicator-based sustainability assessment, the literature contains a considerable number of studies with different concerns, such as development, market and economy, innovation and knowledge, and ecosystems [5]. These studies employ indicators or index-based models to perform sustainability performance evaluation, and are invaluable sources that shed light on the practicability and theoretical strengths of the SILENT Model.

The data requirements of the SILENT Model highlight the dual relationship between theoretical robustness and data accessibility and quality considerations. While the theories related to the variables of urban sustainability considerations convey a very wide and interrelated picture, finding respective data from available sources is not always an easy task. In some cases, available data may not have the desired scope, or have statistical flaws that may result in biased measuring and forecasting. Additionally, auto-correlation between indicators is another issue that could jeopardise model reliability. In some respect, the selection of data is based partly upon intuition and partly on subjective judgement, a situation not uncommon when building a decision support model [5,32]. A carefully designed indicator selection procedure of the model helps in making the model more concise and cost effective by avoiding unnecessary data collection costs. Data availability effects the selection of suitable indicators. As explained by Hak et al. [30], indicators are merely assessment tools; therefore, the cost of improvements should not limit the capacity to implement policy. This is to say as in the case of the SILENT Model, indicators also need to be selected in a cost-effective way.

8.3.1 CONCEPTUAL BASE OF THE MODEL

Based on the aforementioned conceptual and technical issues, the review of the literature and the best practice model and cases, the SILENT Model accommodates four key indicator category areas of "demography" "land use and urban form," "transport" and "environment," in order to best explain the social, economic and environmental sustainability of an urban locality under investigation (see Figure 1). As the literature indicate, these four broad categories represent all of the major human activity areas that play a critical role in effecting sustainability levels of urban environments [2,15]. These categories are considered as generic global categories so as the indicator sets. In terms of indicators this study purposely selected suitable indicators for the case of Gold Coast City, Australia. However, the methodology presented in this paper could be easily replicated elsewhere as long as indicators are carefully selected to reflect sustainability characteristics of the local context clearly.

8.3.2 INDICATOR BASE OF THE MODEL

In keeping pace with the growing interest in sustainability research, there have been various studies proposing different scope and content

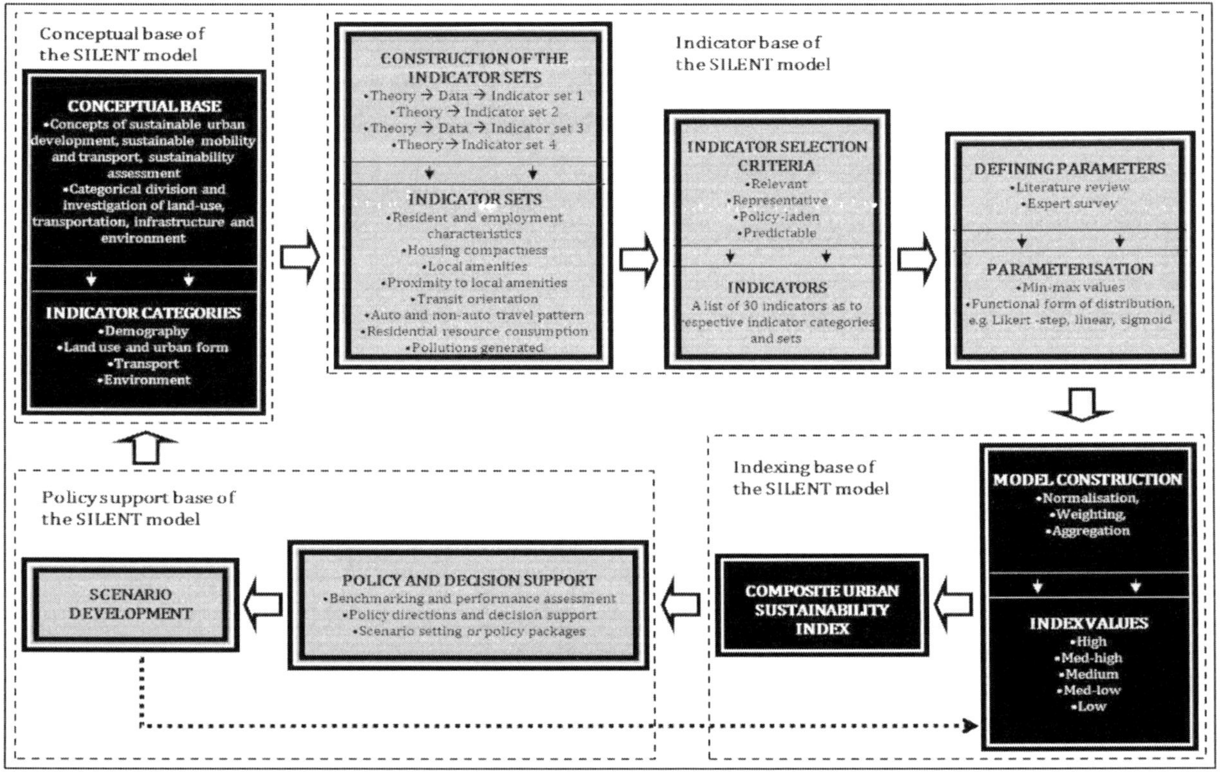

Figure 1. Structure of the SILENT Model.

for sustainability indicators. Also, depending on the scale of the consideration, it is very common to see local and regional indicators defined separately for sustainability assessment. Putting the spatial scale aside, one of the strengths of the SILENT Model is its indicators being conceptually robust and practically relevant to the urban sustainability context. While conceptually robustness refers to the inclusion of all of the key indicators based on the relevant theoretical grounding, the practical relevancy highlights a selection of suitable indicators by considering data availability, policy guidance, and institutional and public priorities. The main difficulty faced while using indicators is to find a common unit of measurement to compare performance of the setting or policy package. Gasparatos et al. [33] state that there are three widely used sustainability assessment methods: monetary tools; biophysical models; and sustainability indicators and composite indices. As done for cost-benefit analysis of environmental assessment, converting externalities into monetary terms is the most preferred approach, while another popular method is to convert parameters into units of global hectares as required by the carbon footprint concept. Biophysical models refer to entropy and carrying capacity concepts. For example, the global hectares concept posed by the carbon foot-printing method is a biophysical measure which is easily understandable, comparable, and frequently used for policy formulation. However, converting some of the social and environmental qualities such as traffic fatalities, endangered species, protected habitats, and human health into common units might be a challenging task. In the light of the literature findings, the SILENT Model uses ecological foot-printing as a common unit of measurement in its sustainability indicators and composite indices-based assessment methods.

As mentioned previously, the SILENT Model uses an indicator-based assessment system. From a pool of over 600 indicators, gathered from a thorough review of the literature [30,34,35], the most relevant indicators, 30 of them, are selected to form the indicator system. The indicator selection decision is made by a team of dozen experts, five researchers, five practitioners and two local government policy makers, in a collaborative manner through a series of workshops. The most relevant indicators for the study area of the Gold Coast City, Australia are selected by considering the literature (theory), local context and data availability (practicality). Each indicator is located under a relevant indicator set, and each indicator set is placed under an indicator category based on their theoretical relationship. The SILENT Model uses suitable parameters in the indicator-based measurement process. Assigning parameters to indicators is not always an easy task. In some cases, it is inevitably hard to define parameters, especially for social and value-dependent measures. For these measures in the model, searching for an innovative approach, localising measures via public involvement and reconciliation, or using proxy variables is considered a solution. If parameters can be determined via literature and other sources, they are used. If not, new parameters are established by using Delphi method. For the Delphi method the same aforementioned expert team of a dozen people are contacted through a set of face-to-face meetings for interviews and surveys in order to determine most suitable parameters for each indicator. The indicator category, indicator sets and individual indicators are listed in Table 1. Additionally Appendix 1 lists a more detailed version of this table including parameters and sustainability calculation formulas of each indicator.

8.3.3 INDEXING BASE OF THE MODEL

In the literature, the terms "composite indicators" and "indices" are generally considered as synonymous [5,36]. While the final product of some studies is a composite indicator, others produce a series of comparable indices. Particularly in measuring sustainability, these are grouped under the usual environmental, economic and social indices [28]. The main characteristic of the indices is that they do not have a unit, so that they provide comparison opportunity between localities [31]. The procedure followed in generation of the indices also points out the main weakness of the composite indicators. Components are assigned weights with the proportion of variances in the original set of indicators, and can then be aggregated using an additional or a functional nature. Weightings are used to correct the information overlap of correlated indicators, so as to ensure that the results are not biased [32]. The weighting methodology carries a value-dependent bias and, in some cases, weighting with linear aggregation causes substitution among indicators. This gives rise to overly-normalised index values

Table 1. Indicator System of the SILENT Model.

Indicator Categories	Indicator Sets	Individual Indicators
Demography	Residential characteristics	Population density
		Labour force participation
		Car ownership
	Employment characteristics	Jobs to housing balance
		Employment density
Land Use and Urban Form	Housing compactness	Mix use ratio
		Dwelling density
		Single-family parcel size
		Single-family dwelling density
		Multifamily dwelling density
	Local amenities	Recreation facility supply
		Socio-cultural facility supply
Transport	Transit orientation	Transit adjacency to residents, services
		Transit patronage ratio
	Non-automobile travel pattern	Transit adjacency to employment
		Transit proximity to employment
		Pedestrian network coverage
		Bicycle network coverage
	Automobile travel pattern	Home-based vehicle kilometres travelled
		Non-home-based vehicle kilometres travelled
		Number of home-based vehicle trips
		Number of non-home-based vehicle trips
		Parking supply in employment centres
Environment	Residential resource consumption	Wastewater generation
		Solid waste generation
		Energy use
		Residential water consumption
	Pollution generated from traffic	Greenhouse emissions generated
		Stormwater runoff pollution generated
		Noise pollution generated

[36]. The SILENT Model determines its indicators' weightings through a Delphi study. Aforementioned expert team is also consulted to help authors with assigning appropriate weightings to each indicator.

However, aggregation of these indicators as an index can cause, in some cases, critical information losses which make it difficult to identify negative or positive changes in the indicator due to the offsetting effects of the positive indicators on negative ones. A good example is from Oregon, US, where a framework measuring environmental, social and economic sustainability levels showed a rise in social and economic indices and a falling environmental index, but with a rise in the overall sustainability index [37,38]. The inability to identify negative movement of indicators may lead to remedial efforts that are applied too late; this would then render the whole exercise fruitless. Composite indices have also been criticised for their inability to show the negative movements of particular indicators, thus making it difficult to implement strategies that target specific problem areas [39]. Therefore, while working with composite indices, the SILENT Model also uses control indicators in a disaggregated form or, at least, to select critical indicators that can be used for early warnings about critical mistakes.

Besides the earlier mentioned Delphi study the SILENT Model also uses a set of statistical methods in order to make sure the most suitable indicators are selected. The statistical analysis is undertaken to clarify the relationship between indicators and urban sustainability by employing Factor Analysis Technique, one of the widely used multivariate analysis methods. Before determining the factors impacting sustainability levels, a correlation matrix is formed to select indicators to be analysed. After this, factors are designated via checking screeplot of eigenvalues, and then indicators are assigned to the respective factors by using rotated loading matrix technique. This process is useful in providing a lesser number, conceptually sound and relatively independent set of factors, where they form the main drivers of urban sustainability within the framework of the SILENT Model.

The second step of the SILENT Model is to normalise the values of each indicator before weighting and aggregation procedures. There are three widely used methods for normalisations [5]. The first method is to use a standardised distribution, such as normal or t-distribution. Secondly, it is possible to convert all values into standard ordinal scale (e.g., Likert scale). Thirdly, linear arithmetic normalisation procedures could be employed using minimum and maximum values of the indicators. The main differences between these approaches are that they give different weightings to the values according to their difference from the mean value. Or, as in the Likert scale, values are placed into distribution-free scale, thus potentially bringing researchers, practitioners and public perceptions together for the normalisation procedure. This study uses the Likert scale in order to convert all values into standard ordinal scale (i.e., Low, Medium-low, Medium, Medium-high, and High). This normalisation method is proven to be useful and used in similar indexing studies [31].

The third step involves assigning the weighting of each indicator or factor. Various techniques such as multivariate analysis of factor analysis, and public and expert opinion techniques are employed for this procedure [5,30,34]. The main consideration at this stage is to select a robust method that evaluates weightings as to their relative importance in the model or, alternatively, in the decision making procedure. The latter consideration is the main reason for the Delphi method.

The fourth step of the model is aggregation of the respective indicators to produce a set of indices and a composite index. While the literature shows us that simple additive rules are generally employed, it is possible to define a functional form for aggregation. As stated by Singh et al. [5], composite indices should ideally remain relatively simple in terms of their construction and interpretation, and the choice of method employed in weighting and aggregation is ultimately dependent on the nature and scope of the particular case study. In the SILENT Model aggregation and disaggregation method is undertaken in two scales. The first one is the aggregation of normalised data from street and parcel levels into grid cells, and also at the same time disaggregation of Census data from Census Collection Districts (CCDs) into grid cells. The second one is aggregation of grid cell sustainability levels into CCDs, postcode areas and suburbs. The last step of the SILENT Model is visualisation of the composite index values in a GIS environment. ArcGIS is used as a system for spatial analysis and visualisation. The GIS-based model produces a grid cell system for sustainability analysis. The study area is divided into 100 × 100 meters grid cells and composite sustainability index values of all indicators are

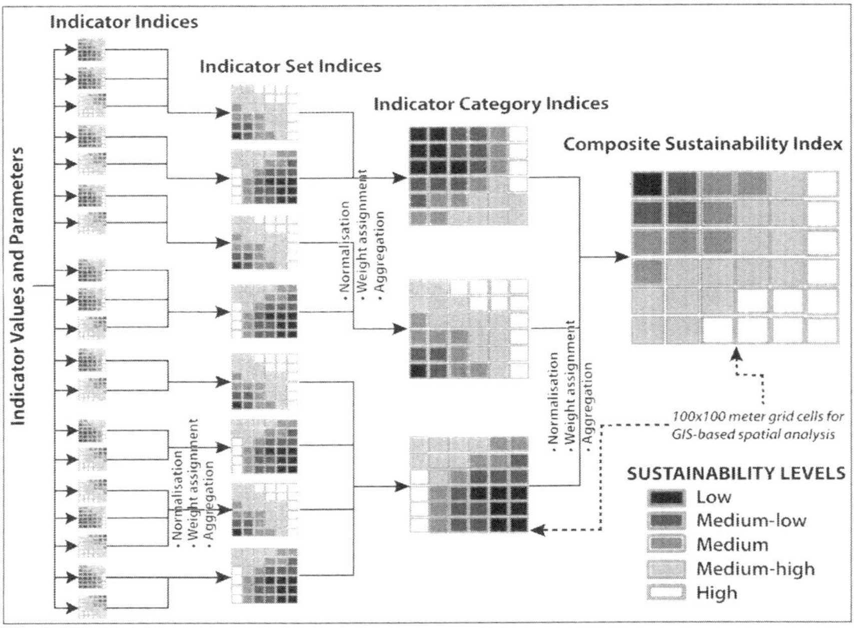

Figure 2. Composite Sustainability Indexing Structure of the SILENT Model.

transferred into the grid cells. Following the entry of the weighting factors, the GIS system produces a set of indices including composite sustainability index in five comparative sustainability levels: Low, Medium-low, Medium, Medium-high, and High. These five comparative sustainability levels are set by assigning calculated Likert scale values for each indicator between the value of 0 and 5: Low (0.00–1.00), Medium-low (1.01–2.00), Medium (2.01–3.00), Medium-high (3.01–4.00), and High (4.01–5.00).

The GIS system also provides a tabular report for exact unit values of the comparative sustainability level of each grid cell. This report contains a sustainability figure each, between 0 and 5, for 30 indicators, a figure each for nine indicator sets, a figure each for four indicator categories, and a composite index value representing the overall sustainability level of this particular grid cell. These grid cell values via the aggregation method in the GIS environment are converted into other geographical scales of analysis, such as street, neighbourhood, CCD, suburb and city [31]. As well as the composite urban sustainability map, a map each is prepared for all indicators, indicator sets, and indicator categories. Figure 2 above illustrates the basic composite urban sustainability indexing structure of the GIS-based model.

8.3.4 POLICY SUPPORT BASE OF THE MODEL

The SILENT Model develops a set of indices and a composite sustainability index to be used for benchmarking and performance assessment of comparative urban sustainability levels and development of relevant policies and strategies, considering both current and future comparative sustainability levels. This allows for the review of the capacity and comparative sustainability levels of current urban formation, and enables the forecasting of future scenarios, via simulation, which local and state governments, planning institutions and firms and local community organisations could highly benefit from.

This indexing model can be used for informing policy, strategy formation and also as a planning or decision support system. Some of the particular planning policy areas that the SILENT Model is relevant to include: Planning and managing sustainable urban development; Planning the development of sustainable transport infrastructure and services; Planning for and prioritising sustainable urban infrastructure; Assessing the development applications; Designating conservation areas; Safeguarding existing environmental assets and values; Developing policies for sustainability and intervening with unsustainable development; Increasing awareness among the community via participatory planning mechanisms promoting urban sustainability.

The SILENT Model has the communicative advantage of being easy to convey comparative levels of sustainability, making it a relatively simple exercise for both the general public and decision makers to understand. The model can also be used for forecasting, with future infrastructure scenarios to be evaluated using predicted data and development trends.

As the SILENT Model is recently developed, at this early stage of the project it could only be tested with dummy figures in a case study in the Gold Coast, Australia. The main purpose of this dummy pilot study is not to measure accurate sustainability levels, but to see whether the model works properly and provides meaningful findings. Figure 3 below demonstrates an example of the composite index developed for part of the Gold Coast City by using hypothetical data. Unsurprisingly the application of the model in a hypothetical exercise in the Gold Coast showed that areas around major arterial roads and main activity centres generally have low sustainability levels compared to those close to green spaces and natural environment. This experiment has demonstrated that the model in the broad sense working properly and ready for minor adjustments and calibration. Once the model is calibrated based on this pilot testing, it will be run with real data.

Authors of this paper and the rest of the research team are commissioned by the Local Government of Gold Coast City, and the State Transport Authority of Queensland Transport and Main Roads, to undertake a case study for the three selected Gold Coast suburbs—Coomera, Helensvale and Nerang—by using the SILENT Model to provide more accurate assessment of their existing and future sustainability levels. The data collection stage of this case study has already been commenced and it is expected to be completed in early 2010, and then the SILENT Model will be applied to the selected suburbs of the Gold Coast.

8.4 CONCLUSIONS

The research results demonstrate that it is possible to produce a viable local level sustainability assessment model, apply the model to a major urban area (e.g., Gold Coast City), and produce a mappable sustainability index. However, this paper only describes the first iteration of the SILENT Model. In this first run we only looked at the basic four key dimensions of urban sustainability (i.e., urban demography, land use and urban form, transport and the environment). Parallel to the views of Gasparatos et al. [40,41], we also acknowledge that additional aspects are also needed to be considered (e.g., equity, participation, and the precautionary principle).

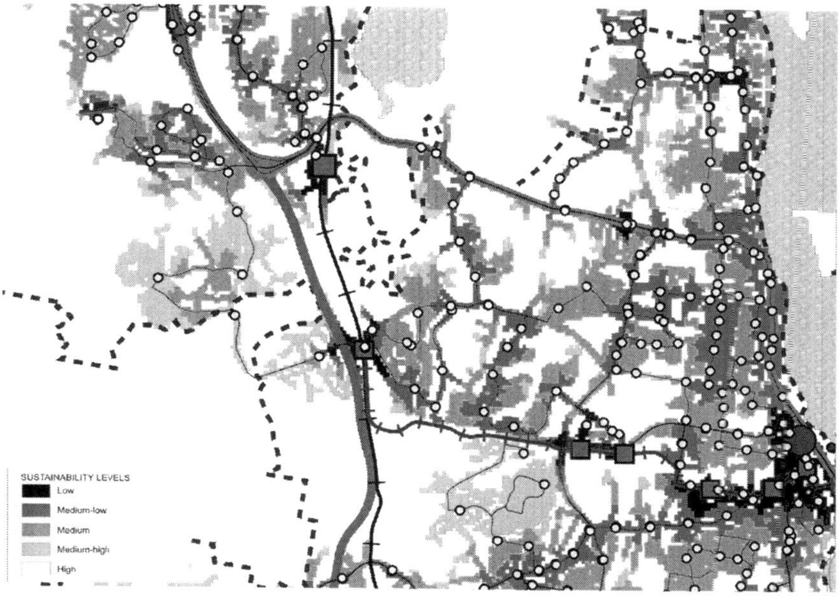

Figure 3. Sample Composite Indexing Map of the SILENT Model

Hence, further research is anticipated which will focus on enhancing the model by testing various indicators in order to best reflect comparative sustainability levels of urban localities. Another area for further development and amendment of the model involves the inclusion of infrastructures other than transport (e.g., water, sewerage, stormwater, power) in to the SILENT Model. Moreover, decreasing the grid cell sizes and developing a parcel-based module of the model are among the improvements to be explored in future refinements. All these improvements will also be tested in several pilot studies, and several sensitivity analyses with different weightings will be conducted before the model is potentially adopted into Gold Coast City Council's planning mechanism.

While still in its infancy, the SILENT Model has been tested on the Gold Coast case study by using hypothetical data. The sustainability indexing and assessment experience has shown that, when fully operationalised, the model has the potential to help planners and policy makers to pursue an integrated framework for locally adoptable sustainability policies. The model is useful in providing unambiguous representation of relationships in urban form and problem areas of urban settings and, where necessary, policies can be tested and accommodated. The model employs a holistic view of urban dynamics and is not only an invaluable sustainability and environmental impact assessment model, but also a practical planning decision support system. When considered in the context of growing population, urban and environmental problems and climate change, the SILENT Model has an immense potential to aid involved parties in forming sustainable urban and transport development policies and in monitoring their impacts on the environment.

REFERENCES

1. Yigitcanlar, T.; Fabian, L.; Coiacetto, E. Challenges to urban transport sustainability and smart transport in a tourist city: The Gold Coast, Australia. Open Transport. J. 2008, 1, 19-36.
2. Teriman, S.; Yigitcanlar, T.; Mayere, S. Urban sustainability and growth management in South-East Asian city-regions: The case of Kuala Lumpur and Hong Kong. Plann. Malays. 2009, 7, 47-68.
3. Hemphill, L.; Berry, J.; McGreal, S. An indicator-based approach to measuring sustainable urban regeneration performance. Urban Stud. 2004, 41, 725-755.
4. Hezri, A.A. Utilisation of sustainability indicators and impact through policy learning in the Malaysian policy process. J. Environ. Assess. Pol. Manag. 2005, 7, 575-595.
5. Singh, R.K.; Murty, H.R.; Gupta, S.K.; Dikshit, A.K. An overview of sustainability assessment methodologies. Ecol. Indicat. 2009, 9, 189-212.
6. Brownhill, D.; Rao, S. A Sustainability Checklist for Developments: A Common Framework for Developers and Local Authorities; BRE Press: London, UK, 2002.
7. Europe's Environment: The Dobris Assessment; Stanners, D., Bourdeau, P., Eds.; European Environment Agency: Copenhagen, Denmark, 1995.
8. Brandon, P. Foreword. In Sustainable Urban Development: The Framework and Protocols for Environmental Assessment; Curwell, S., Deakin, M., Symes, M., Eds.; Routledge: New York, NY, USA, 2002.
9. Indicators of Sustainable Development: Guidelines and Methodologies; United Nations Publications: New York, NY, USA, 2007.
10. Environmental Indicators; Organisation for Economic Cooperation and Development Press: Paris, France, 1994.

11. Impact Assessment and Sustainable Development: European Practice and Experience; George, C., Kirkpatrick, C., Eds.; Edward Elgar Publishing: London, UK, 2007.
12. Foley, B.; Daniell, A.; Trevor, M.A. Sustainability tool for intrasectoral and intersectoral water resources decision making. Aust. J. Water Resour. 2004, 8, 11-19.
13. Norman, W.; MacDonald, C. Getting to the bottom of triple bottom line. Bus Ethics Q. 2004, 14, 243-262.
14. Karol, P.; Brunner, J. Tools for measuring progress towards sustainable neighborhood environments. Sustainability 2009, 1, 612-627.
15. Hacking, T.; Guthrie, P. A framework for clarifying the meaning of triple bottom-line: Integrated, and sustainability assessment. Environ. Impact Assess. Rev. 2008, 28, 73-89.
16. Shepherd, A.; Ortolano, L. Strategic environmental assessment for sustainable urban development. Environ. Impact Assess. Rev. 1996, 16, 321-335.
17. Dalal-Clayton, B.; Sadler, B. Strategic Environmental Assessment: A Sourcebook and Reference Guide to International Experience; Earthscan: London, UK, 2005.
18. Handbook on Constructing Composite Indicators: Methodology and User Guide; Organisation for Economic Co-operation and Development (OECD): Paris, France, 2008.
19. Meadows, D. Indicators and Information Systems for Sustainable Development; The Sustainability Institute: Hartland, WI, USA, 1998.
20. Warhurst, A. Sustainability Indicators and Sustainability Performance Management; International Institute for Environment and Development: Warwick, UK, 2003.
21. Wackernagel, M.; Rees, W.E. Our Ecological Footprint: Reducing Human Impact on the Earth; New Society Publishers: Philadelphia, PA, USA, 1996.
22. Wackernagel, M.; Kitzes, J.; Moran, D.; Goldfinger, S.; Thomas, M. The ecological footprint of cities and regions: Comparing resource availability with resource demand. Environ. Urban. 2006, 18, 103-112.
23. McManus, P.; Halighton, G. Planning with ecological footprints: A sympathetic critique of theory and practice. Environ. Urban. 2006, 18, 113-127.
24. Atkisson, A. Developing indicators of sustainable community: lessons from sustainable Seattle. Environ. Impact Assess. Rev. 1996, 16, 337-350.
25. Atkisson, A.; Hatcher, R.L. The compass index of sustainability: Prototype for a comprehensive sustainability information system. J. Environ. Assess. Pol. Manag. 2001, 3, 509-532.
26. Valentin, A.; Spangenberg, J.H. A guide to community sustainability indicators. Environ. Impact Assess. Rev. 2000, 20, 381-392.
27. Astleithner, F.; Hamedinger, A.; Holman, N.; Rydin, Y. Institutions and indicators: The discourse about indicators in the context of sustainability. J. Hous. Built. Environ. 2004, 19, 7-24.
28. Lautso, K.; Spiekermann, K.; Wegener, M. Modelling policies for urban sustainability. In Proceedings of the 42nd Congress of the European Regional Science Association, Dortmund, Germany, August 2002.
29. Maclaren, V.W. Urban sustainability reporting. J. Am. Plann. Assoc. 1996, 62, 184-202.

30. Hak, T.; Moldan, B.; Dahl, A.L. Sustainability Indicators: A Scientific Assessment; Island Press: Washington, DC, USA, 2007.
31. Yigitcanlar, T.; Sipe, N.; Evans, R.; Pitot, M. A GIS-based land use and public transport accessibility indexing model. Aust. Plan. 2007, 44, 30-37.
32. Hanafizadeh, M.R.; Saghaei, A.; Hanafizadeh, P. An index for cross-country analysis of ICT infrastructure and access. Telecomm. Pol. 2009, 33, 385-405.
33. Gasparatos, A.; El-Haram, M.; Horner, M. The argument against a reductionist approach for assessing sustainability. In Proceedings of International Conference on Whole Life Urban Sustainability and Its Assessment, Glasgow, UK, June 2007.
34. Hass, J.L.; Brunvoll, F.; Hoie, H. Overview of Sustainable Development Indicators Used by National and International Agencies; Organisation for Economic Co-operation and Development (OECD): Paris, France, February 2002.
35. Towards a Local Sustainability Profile: European Common Indicators; Technical Report for European Commission: Brussels, Belgium, 2001.
36. Munda, G. Multiple criteria decision analysis and sustainable development. In Multiple Criteria Decision Analysis: State of the Art Surveys; Figueira, J., Greco S., Ehrogott, M., Eds.; Springer: New York, NY, USA, 2005; pp. 953-986.
37. Schlossberg, M.; Zimmerman, A. Developing state-wide indices of environmental, economic, and social sustainability: A look at Oregon and the Oregon Benchmarks. Local Environ. 2003, 8, 641-660.
38. Frame, B.; Vale, R. Increasing uptake of low impact urban design and development: The role of sustainability assessment systems. Local Environ. 2006, 11, 287-306.
39. Neuman, M. Infiltrating infrastructures: On the nature of networked infrastructure. J. Urban Technol. 2006, 13, 3-31.
40. Gasparatos, A.; El-Haram, M.; Horner, M. The argument against a reductionist approach for measuring the progress towards sustainability. Account. Forum 2009, 33, 245-256.
41. Gasparatos, A.; El-Haram, M.; Horner, M. A critical review of reductionist approaches for measuring the progress towards sustainability. Environ. Impact Assess. Rev. 2008, 28, 286-311.

CHAPTER 9

Quantifying the Total Cost of Infrastructure to Enable Environmentally Preferable Decisions: The Case of Urban Roadway Design

CONRAD A. GOSSE AND ANDRES F. CLARENS

9.1 INTRODUCTION

Ground transportation is responsible for nearly 30% of the primary energy consumption and 27% of the greenhouse gas (GHG) emissions in the United States [1]. Related infrastructure also results in significant material movement—every $1 million investment in roadway construction requires 9×10^4 tonnes of aggregate and 3.3×10^3 tonnes of cement [2]. In an effort to try to reduce these burdens, numerous life cycle assessments (LCAs) have been performed over the past decade to understand how specific technological choices contribute to emissions, energy consumption, and materials use. These studies have tended to focus on either the design or the use phase of the road [3]. Design generally involves the selection of a material, e.g., concrete or asphalt, or specification of roadway width and configuration [4]. Use entails a variety of other processes including vehicle selection or roadway maintenance [5]. Even though these studies have identified many obvious opportunities for environmental improvement, there is little evidence to suggest that they have provided deep reductions in material use or emissions.

A principal limitation of many published analyses is that they consider technological options for ground transportation narrowly and evaluate only specific elements of the design or use of roadways at once. An asphalt road may have a lower life cycle emission profile than a concrete road, for example, but that difference is small when compared to the overall emissions from the use phase of the road [6]. Similarly, the conclusion that a greater bicycle mode share will reduce the carbon emissions of a roadway is not useful if it is not considered along with other factors discouraging bicycle use and the impact of more bicycles on overall traffic flow. To date, LCA has been employed as a method for environmental bean counting that considers problems removed from the broader system within which they exist. Consequently, even though LCA has been actively pursued in academic circles, it has had only limited impact in policy circles.

At the same time that conventional LCA tools have been insufficient for solving many of the existing problems faced by transportation managers, emergent challenges make the need for new tools even more pressing [7]. Conventional development patterns have led to widespread congestion in urban and suburban areas around the world. Shrinking maintenance budgets at a time when facilities built in the post WWII boom period are reaching their design life span are making it ever harder to maintain the level of service that was envisioned for roadways during design. Declining pavement quality is also exacerbating the emissions [8] and safety costs associated with the use of these aging facilities, which only compound the impacts of unchecked growth in vehicle kilometers of travel (VKT) worldwide. Efforts to address these problems and provide meaningful improvements will require systems thinking that considers life cycle impacts, personal choice, and policy realities.

In practice, ground transportation is constrained by a few overarching factors. The most obvious is budget. Road construction is expensive, though less so than alternatives like public transport, because much of the cost is borne by users in the form of vehicles and fuel. Roadways are expensive to maintain, and so many exhibit condition ratings below their design values. In many urban areas, space is also constrained and a limiting factor in terms of enabling more mobility. Where space is available, additional road and parking facilities relieve congestion in the near term but only further separate typical destinations, increasing trip lengths over the long term [9]. The carbon emissions from ground transportation are appreciable and growing as more and more developed nations move toward car ownership levels on par with the United States. Paradoxically, the convenience afforded by automobiles has contributed to significant and consistent traffic-related delays in almost all of the worlds' urban centers. These delays translate into appreciable costs to the users of the transportation systems [10], in addition to increased crash and health risks [11], and overall environmental impacts.

Planning that tackles these challenges involves both near-term adaptive strategies and long-term improvement projects. Existing facilities, including the functionally obsolete, have significant embodied emissions and sunk costs that preclude their immediate replacement, even when sufficient funds are available. This lag between identifying changing needs and building new infrastructure results in increased total public costs with respect to design projections. Interim adaptive strategies can be considered, however, to minimize the monetary, environmental, and safety impacts of a sub-optimal design still in the middle of its useful life until it is time to replace the facility [12]. Adapting existing facilities to new use patterns and goals also provides a bridge between generational shifts in infrastructure planning objectives.

In the case of transportation, planning has historically focused on mobility, with the outcomes of ever-increasing VKT and sprawling development patterns that discourage alternatives to the private automobile [13, 14]. In the near term, however, adapting existing facilities to maximize mobility in light of an increased bicycle and transit mode share is an appropriate measure to maximize the value provided by these facilities. Adaptive strategies comply with existing system constraints and involve lower cost measures to capture the remaining value of past infrastructure investments until such time as a major change consistent with a lower impact vision is warranted. Evaluation of these transitional actions is complicated, however, given the absence of any sort of steady state and the number of analyses that must be integrated.

Most of the individual elements constraining ground transportation systems have been studied in isolation, but few examples of integrated multi-criterion analysis of roadway use have been published. Pavement management systems (PMS) have been developed to help maintenance managers maintain large systems of pavements under budget constraints [15–18]. Separately, traffic engineers have developed microsimulation tools of vehicle dynamics to understand the effect of different road configurations [19–21] and work-zone traffic management [10]. Economic analyses rely in part on the engineering analysis of road utilization and make the connection between more infrastructure and induced economic activity. The interests of pedestrians and bikers are also considered in the context of safety [22–24] and congestion [25] but rarely in terms of providing viable alternatives to automobile transportation.

Efforts to identify significant reductions in the environmental burden of transportation will need to consider these tools together to inform optimal use of roadways under multiple constraints. Here we present a method for combining these analyses with conventional LCA of roadways. We consider the results in the context of adaptive roadway lane (re)configuration, such as converting curb parking to bicycle facilities, that many cities, including Washington, DC and New York City, are currently undertaking in an effort to reduce congestion directly and indirectly by supporting alternatives to automobile travel that make more efficient use of the public right of way and incentivize reduced trip distances.

9.2 METHODS

A total cost minimization approach is proposed in order to identify preferable lane configurations for two-lane urban roadways, given the physical parameters of the site, available right of way width, and traffic volumes by mode. A lane configuration is defined by the number and width (or presence) of parking, bicycle, and conventional travel lanes for each direction. Costs include: annualized pavement maintenance, motor vehicle fuel costs [26], and travel time at half the prevailing wage rate [27]. GHG emissions are also calculated. Computationally, the proposed framework is structured as a series of distinct codes.

Microsimulation of idealized roadway segments was carried out using VISSIM 5.4 for all parameter combinations given in table 1 using common values in table 2. The first section of table 1 lists the parameters that define a lane configuration. The remaining parameters in table 1, taken together, will be referred to as the scenario. The key dynamic explored in this work is the use of a single lane by motor vehicles and bicycles simultaneously, and whether the former is able to safely pass the latter within the lane. VISSIM is able to model lateral behavior within lanes, in addition to more conventional vehicle following and lane changing behaviors, and so is able to consider this question. Peak and off peak traffic volumes were simulated separately and combined using 12 h of each to arrive at daily totals, which were then inflated to annual values. Additional detail is provided in the supporting information (available at stacks.iop.org/ERL/8/015028/mmedia).

A low and high bicycle mode share were explored, with 1% representing typical urban mode share in the US and 10% representing a target that leading cities, such as Portland, OR, could achieve in the next decade with sufficient investment. Total person trips are constant between the bicycle volume scenarios, with the difference made up by automobiles with an occupancy of 1.2. Figure 1 shows the relative sizes of vehicles and lanes considered in this work. The design standard for lane width in the US is 3.6 m [28], however it is both permissible and likely that narrower lanes are used in the width constrained urban corridors considered in this study, so we assume a 3.4 m base case. Assuming a 1 m passing buffer, 1.75 m wide automobile, and 0.5 m wide bicycle, an automobile is able to pass a bicycle within the lane, however a 2.6 m wide transit bus is not. To evaluate the benefit of alleviating this 'stuck' condition, a wider lane is also considered that allows buses to pass cyclists without either vehicle departing the lane. 'Dooring' accidents, crashes between bicycles and opening automobile doors, may also justify additional buffer width between parked cars and lanes used by bicycles, however the present work does not model crashes, so this effect is not represented in the analysis. The impact of these passing conditions is dependent upon the likelihood that buses will

Table 1. Discrete segment parameter space. All unique combinations were evaluated using microsimulation. The first group defines a lane configuration and the second a scenario.

Travel lane width	3.4 m, 4.3 m
Bicycle lane	None, 1.22 m[a]
Parking lane	None, 2.5
Characteristic length between passing zones	50 m, 100 m, 200 m
Grade	0%, ± 4%
Bicycle mode share	1%, 10%

[a] AASHTO guidelines call for a wider 1.52 m bicycle lane adjacent to curb parking which are applied here as appropriate.

Table 2. Common parameters and descriptive values for all simulated configurations.

Total width	6.8–16.6 m
Motorized speed	16.7 m s $^{-1}$
Bicycle speed	Varies by grade
Car occupancy	1.2
Bus occupancy	20
Bicycle occupancy	1.0
Potential curb parking	6.1 m spaces covering 80% of segment length

encounter a bicycle and become stuck behind it, thereby delaying themselves and following motor vehicles.

Realistic urban corridors vary in width, so it is overly conservative to assume that a given lane width will restrict passing movements indefinitely. Here, we adopt the concept of a characteristic length between passing zones, as given in table 1 to determine the likelihood that a bicycle and bus will be present and the delay expected to result from the encounter. VISSIM simulates these interactions directly. Since Poisson vehicle arrivals are assumed, however, in order to make the results as general as

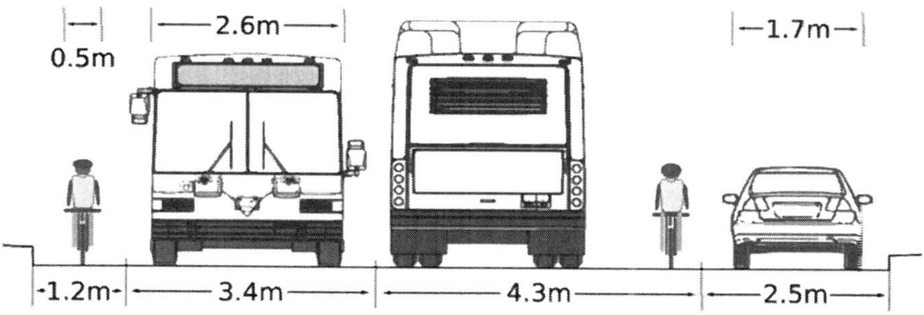

Figure 1. A reference multi-modal street section and the relative widths of vehicles and lanes.

possible, a probabilistic analysis can be carried out to compute expected delay. This analysis is presented in the supporting information (available at stacks.iop.org/ERL/8/015028/mmedia) and agrees with the results of the microsimulation. Motor vehicle speeds were assumed constant for each lane width given the considerable variation that exists in the literature on the effect of lane width of motorist speed choice [22, 29, 28], however, this behavior could be readily altered in the microsimulation parameters where local data is available. Bicycle speeds for each grade were computed according to first principles formulas [30].

Vehicle fuel use and emissions are affected by pavement roughness though not consistently between various operational regimes [31–33] due to the varying contribution of rolling resistance to required power. A power-based vehicle emissions model, CMEM [34], was used to post-process the microsimulation vehicle data at 1 Hz and two roughnesses using a lookup table computed at a reference international roughness index (IRI) of 1.0 m km^{-1}, and for a rough case with an IRI of 4.0 m km^{-1} by inflating vehicle rolling resistance after Karlsson et al [33]. Only automobile emissions were affected due to the inconclusive results of that study for heavy vehicles. Final emissions and fuel consumption were computed by the pavement management module through linear interpolation of the two roughness cases.

Pavement management activity and emissions were computed based on the previous work of the authors [35] with additional dynamic pavement loading due to roughness [36]. Explicit treatment of heavy vehicles is important in the comprehensive analysis of a roadway given their disproportionate impact [37]. Pavement maintenance plans were computed for both directions of travel lanes, bicycle lanes, and parking lanes independently using aggregated annual vehicle volumes and emissions from the appropriate microsimulation trials. A network Pareto front for all the lanes was then computed, with a representative example given in figure 2. For this analysis, the non-dominated plan with the minimum total GHG emissions, subject to agency constraints, was selected. Further detail on this method, and the larger issues of discounting and temporal variation, can be found in the supporting information (available at stacks.iop.org/ERL/8/015028/mmedia).

In combining the separate models of pavement management, vehicle microsimulation, and vehicle emissions computation, a hierarchy exists according to the sensitivity of one to another. Pavement roughness influences vehicle fuel use and emissions as well as dynamic pavement loading. Loading affects pavement durability, and maintenance investment determines the resulting pavement condition. Vehicle behavior is assumed to be insensitive to pavement condition within the specified limits, however, which makes travel time cost insensitive to changes in fuel and agency costs. Critically, this means that the PMS optimization can be performed after microsimulation. Otherwise, the task would be computationally intractable with existing microsimulation tools, since a microsimulation would have to be run at each iteration of the PMS genetic algorithm.

A Pareto front of lane configurations can be identified for each scenario that is non-dominated with respect to total costs and width. Lane configurations that include curb parking will naturally not appear in this set since they incur pavement maintenance costs but provide no counterbalancing benefit as computed. A spatial opportunity cost of curb parking can be computed, however, by computing the cost differential between configurations with parking and a point interpolated on the Pareto cost curve at the same total width. This opportunity cost allows decision makers to quantify the potential mobility value of public right of way allocated to parking.

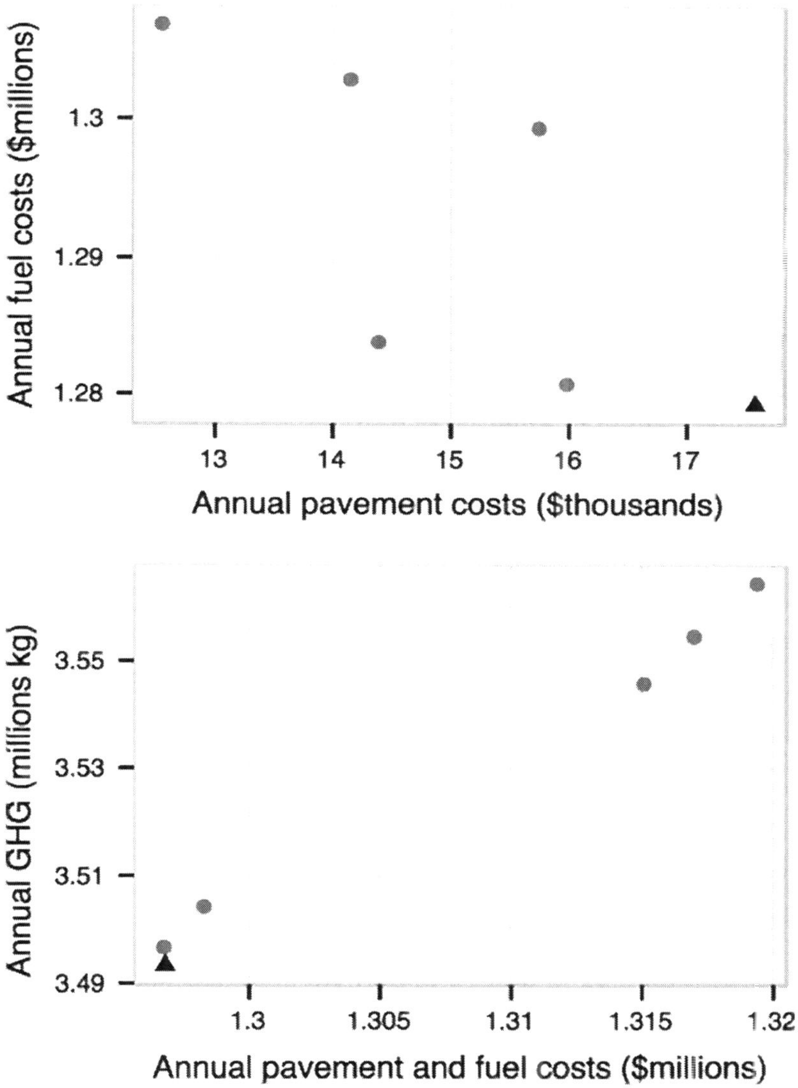

Figure 2. Agency pavement costs versus user fuel costs as a result of pavement roughness (upper) and combined costs versus combined GHG emissions (lower) from non-dominated pavement management plans. The triangle data point indicates the selected plan by minimum GHG.

9.3 RESULTS

The effect of bicycle mode share on average travel time for a particular segment can be significant for specific conditions as shown by the results in figure 3. Differentiation between the cases occurs when heavy (wide) vehicles, such as transit buses encounter a bicycle and have insufficient room to safely pass resulting in significant delays for themselves and the motor vehicles behind them. In these graphs, individual data points represent microsimulations of specific cases and are plotted with random jitter on the length axis for legibility. As with other results presented here, they are normalized to a kilometer of travel. The trend lines are second order polynomials used to illustrate the relationships of interest. The effect of grade, and whether or not vehicles are traveling uphill or down, have important effects on the results presented in figure 3. The data are grouped based on whether or not buses are stuck behind bikers in the different configurations. For the level ground segment, the results are equivalent. For the inclined segment, the uphill travel time is always considerably higher than downhill travel time if trucks get stuck behind buses. The impact of the stuck condition is proportional to the relative speed difference between bicycles and motor vehicles, which comes from roadway grade, and the likelihood of a heavy vehicle encountering a bicycle within the characteristic distance between passing zones. This is determined by modal volumes and headway distribution.

For lane widths more narrow than those considered here, all motor vehicles with more than two wheels would be unable to pass bicycles within their lane, with the result that the expected speed of all traffic would approach that of bicycles as characteristic length and bicycle volume increased. These cases are not presented in order to focus on the more typical but less intuitive stuck condition, and because very narrow lanes are likely to have an effect on driver speed decisions according to the particular characteristics of the site, such as sight distance, land use, number of driveways, and other factors. This is not to say, however, that the framework presented here is not suitable for 3.1 m lane widths, only that the results would not be transferable to other situations. For wider lane widths, the differentiation observed here disappears as heavy vehicles are

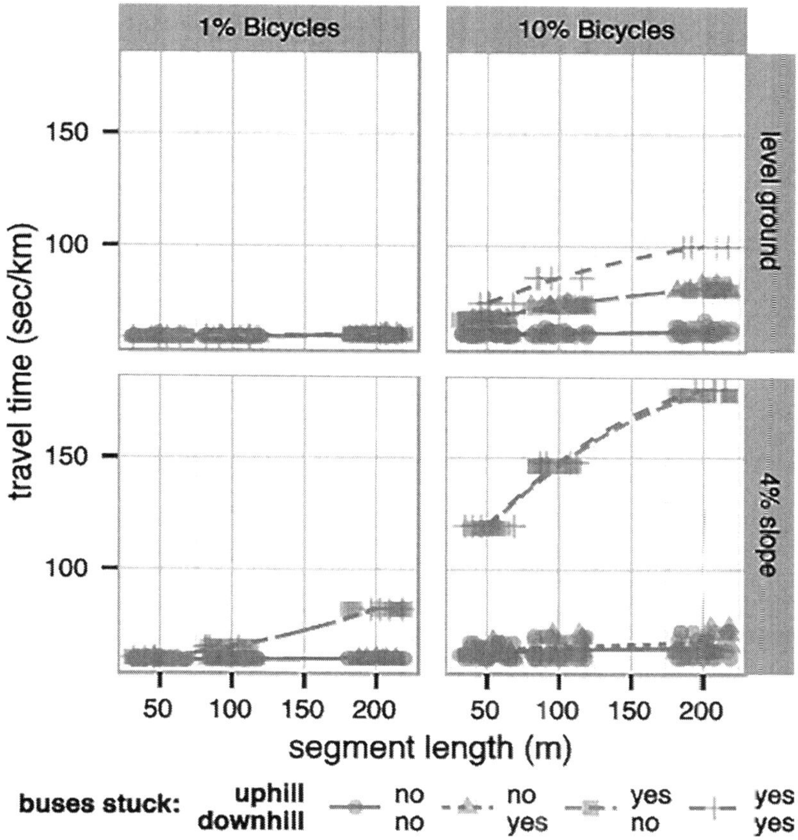

Figure 3. Simulated motorized travel time illustrates the delay caused when heavy vehicles are unable to pass bicycles.

able to pass bicycles. All four stuck groups include multiple lane configurations which reveal more subtle differences in annual costs with respect to width (distance between curbs) when considered individually. In figure 4, this relationship is seen as a distinct Pareto optimal frontier for each characteristic length between passing zones, grade, and bicycle mode share considered, with non-dominated lane configurations shown in bold. The influence of the stuck condition can be seen in the abrupt transition in the Pareto curve around 8.6 m, or the minimum width of a configuration not stuck in both directions. The initial drop between the first two non-dominated points is larger for the 4% case, as compared to level ground, since alleviating the stuck condition on the uphill segment is considerably more important than in the downhill direction given the dramatic difference in expected bicycle speeds.

For the urban situations considered in this work, both travel time and fuel use are positively correlated with vehicle delay. These costs are also considerably larger than agency expenditures for non-dominated pavement maintenance plans, which can be seen in figure 2. This relationship supports the decision to select the PMS plan with the minimum total GHG emissions (higher agency cost), since a comparatively small agency investment provides a larger reduction in user costs. It follows that minimizing total costs also minimizes GHG emissions since total costs are dominated by time and fuel and are sensitive only to delay once pavement condition has been established. This is encouraging since typical planning processes do not explicitly quantify GHG emissions.

Figure 4 presents the Pareto front with respect to total costs and then using delay rather than total travel time costs. Neither approach is strictly more accurate, however since bicycle travel is generally more time consuming over the same roadway segment, increasing bicycle mode share tends to dramatically increase time costs. Using the time cost of delay only, assumes that travelers had already accounted for this cost externally, which is not unreasonable, and is typical in traffic analysis. Under this assumption, figure 4 reveals a tipping point where increased bicycle mode share lowers total costs, given sufficient roadway width. For cases where significant increases in bike ridership are not paired with enhanced facilities like wider roads, this will increase travel time for all users.

Parking is not valued in the total cost reported in figure 4. As a result, lane configurations with parking bays incur pavement maintenance costs without corresponding negative costs from parking's value as a service, rendering these configurations sub-optimal. This also explains why costs reach their minimum in figure 4 approximately 5 m before the maximum width since that is the width occupied by two 2.5 m parking bays. Curb parking does have a site-specific value both in terms of vehicular accessibility and in the broader economic sense by supporting value capture

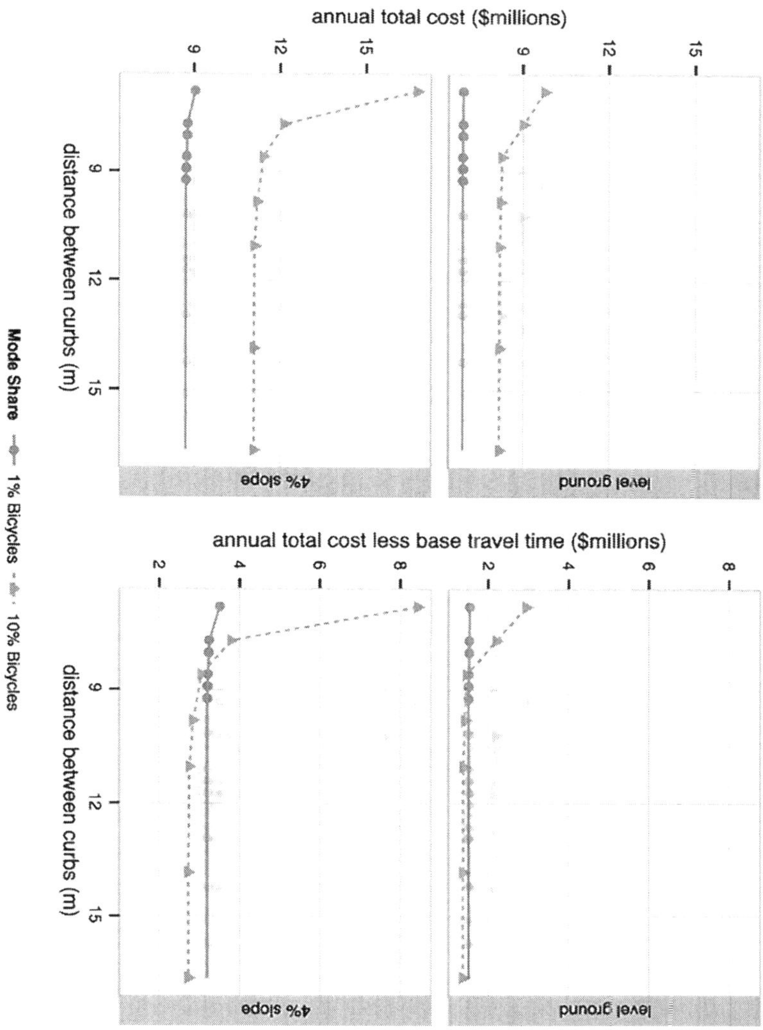

Figure 4. The relationship between road width and costs follows a Pareto optimal behavior, shown here for 100 m characteristic lengths. Total travel time costs (left) dominate fuel and PMS costs, however when delay costs are used (right) increased cycle mode share can reduce costs for appropriate lane configurations.

along the street through increased business patronage. Because the magnitude of this value capture is so site-specific, figure 5 presents a spatial opportunity cost of curb parking as the difference between the total cost of a lane configuration that includes parking and a linear interpolation of the Pareto front at the same width. Visually, this is the vertical distance between the lighter data points in figure 4 and the Pareto front, normalized to a daily value per parking space. The narrowest lane configuration that can include parking is 9.3 m wide, so opportunity cost is reported from this value up to the maximum configuration width.

A spatial opportunity cost of parking is shown in figure 5 for the inclined case with a 100 m characteristic distance between segments. The likelihood of buses' getting stuck behind a bicycle is related to characteristic length, so other scenarios look similar to this plot but shifted in magnitude accordingly by length. Level ground also exhibits consistent behavior, albeit with a different curve shape that can be inferred from figure 4. Since the maximum hourly volumes considered in this study are below saturation levels, delay comes almost entirely from the bicycle/motor vehicle interactions. Therefore, it is not surprising that parking opportunity costs are highly sensitive to the percentage of bicycles in the traffic stream. For a given width, parking and bicycle infrastructure essentially compete for the same space which results in large opportunity costs for parking in narrow configurations, as might be seen in a traditional urban neighborhood of places like New York City and Washington, DC. These larger values are also well in excess of typical parking meter returns, which is consistent with the pervasive subsidization of automobile parking in the US [38]. These values must also be considered conservative since they do not include any opportunity costs for land in the pedestrian zone on either side of the roadway proper that could be put to other uses, such as pedestrian mobility or restaurant seating.

9.4 SENSITIVITY AND ADDITIONAL CONSIDERATIONS

To develop a more complete understanding of how urban roadway design can be informed by a model like this, it is useful to consider both those factors that had little impact on the results and those factors that were not

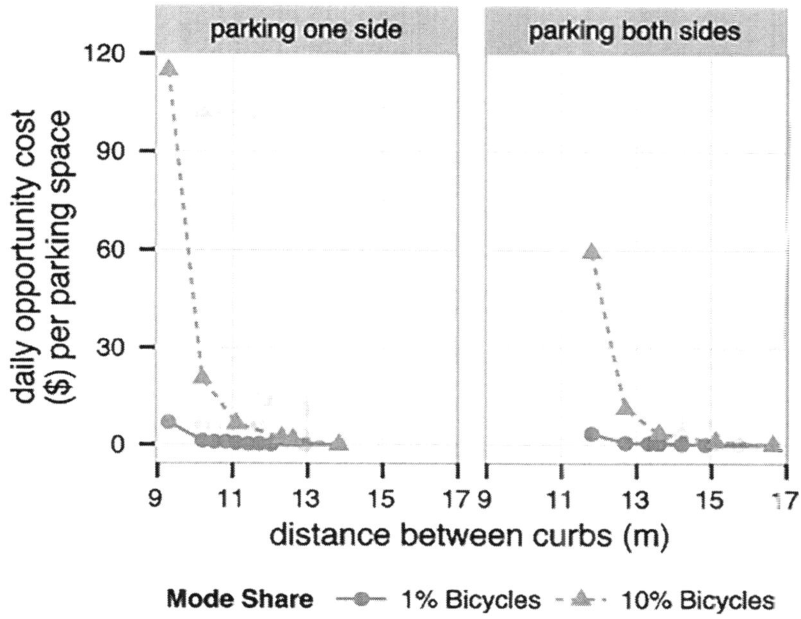

Figure 5. Spatial opportunity cost for curb parking, shown here for 100 m characteristic length and 4% grade, increases dramatically for narrower roads suggesting that when all the costs are considered, there is a tipping point beyond which curb parking becomes an expensive use of land.

included in the analysis for one reason or another. Pavement roughness and its effects on fuel consumption are considered here even though in the context of urban driving, roughness is far less important than acceleration cycles in determining fuel efficiency. However, since travel time is independent of roughness, pavement management costs for a given configuration are optimized against marginal vehicle fuel consumption which is on the same order of magnitude. Additionally, the fiscal reality of the pavement manager may be such that maintaining serviceable pavement is a constant struggle, and so minimization of agency costs may replace minimization of GHG emissions in selecting the best plan. The impact pavement condition has on a given commuter's likelihood to select the bicycle or automobile mode is difficult to quantify, but also more significant to the ultimate makeup of the Pareto set of lane configurations by determining relative volumes between the modes as can be seen in figures 4 and 5.

Similarly, only select measures of the health and safety factors associated with modal shift are captured by this model since many of these are difficult to quantify with confidence even though they are often cited as an important driver and/or obstacle in mode shift toward more active forms of transportation [39]. Crashes, though not considered in this work, are an important factor in selecting bicycle transport and even though AASHTO guidelines dictate an additional 0.3 m of width for bicycle lanes adjacent to curb parking, that width remains insufficient to prevent dooring crashes between bicycles and parked cars [40, 41]. The cost of curbside parking goes up fractionally for this additional 0.3 m of pavement to address the risk of dooring to bikers, however the actuarial cost of a potential fatality is on the same order of magnitude as total annual travel cost per kilometer in this study. Crashes are difficult to model generally, and difficult to even estimate based on past reports involving bicycles due to pervasive under-reporting of non-fatal encounters. Never the less, the potential health and safety costs are potentially large enough to influence the results presented here.

Another limitation of the model is that it does not consider roadway segments within the larger roadway network. In many urban areas, vehicle arrival is not Poisson distributed but rather appears as a decaying platoon progressing between controlled intersections. In addition, driveways play an important roll in platoon migration and overall capacity in urban settings. Finally, buses stop periodically for passengers and this presents an opportunity for the traffic to clear and the biker and bus to separate. These factors were not considered here in order to make the analysis as general as possible but could be readily incorporated into the microsimulation for a specific site. Many of these factors would need to be included in a similar model to derive site-specific estimates for the opportunity cost of parking even though we expect that the general trends discussed here would hold for most urban roadways.

9.5 IMPLICATIONS

A major challenge for urban areas around the world is to improve livability, which is often achieved by reducing reliance on automobile transport [42, 43]. A shift away from the automobile also results in significant reductions in energy use and GHG emissions, both of which are increasingly relevant policy objectives. For countries with a legacy of automobility-dominated planning and policy decisions, such as the US, these goals are especially daunting. Reducing VKT through more compact development and alternative transportation modes is a long-term objective, but previous studies have observed that many of our projected future emissions are 'locked in' by virtue of the inefficient nature of our existing infrastructure stock and the relatively slow rate at which this is replaced [44]. Short-term strategies are needed to achieve some of these gains without major changes to our existing infrastructure. Improving bicycle facilities is one such change that could encourage higher mode share without significant investment. In cities like Washington, DC and New York City, where many of the trips are short and well suited to bicycle transport, efforts have been underway for several years to provide such facilities. Consequently, bike mode share is increasing significantly [45], but the ridership rates in these US cities are still an order of magnitude lower than in many European cities [46]. Additionally, these changes are not without controversy in light of their direct costs and use of limited space [47].

Tools like the one developed here can be used to help resolve the apparent conflicts that could inhibit progress toward more environmentally sustainable infrastructure. We find that increasing bicycle mode share can have a significant impact on motor vehicle delay, and indeed greatly increase total costs where sufficient right of way is not provided. Conversely, with sufficient space to allow wide vehicles to pass bicycles, a reduction in total costs for all users is obtained through an increased bicycle mode share. Parking, specifically curb parking, emerges as space that is potentially most eligible for reallocation to bicycles in width constrained urban corridors because of its ideal position and the relatively minimal expense of lane reconfiguration. The spatial opportunity cost of parking quantifies this tradeoff and exhibits sharp transitions between realistic monetary amounts for wider streets and unfeasibly high expected returns for more narrow areas. These results provide a clear guide to traffic engineers and urban policy makers with respect to optimal allocation of limited pavement.

Even though life cycle assessment and other tools are useful for beginning to understand pieces of this problem, a systems-based approach like the one proposed here is needed to directly support policy decisions. By considering infrastructure systems in an integrated, yet quantitative manner using existing modeling frameworks, short-term low cost opportunities emerge for efficiency improvements that would not be obvious using other tools alone. The general framework proposed here could be applied in a variety of infrastructure contexts. For example, the green building industry standard, leadership in energy and environmental design (LEED), has already recognized that buildings cannot be analyzed outside of the larger infrastructure context within which they exist and is making changes to consider the community within which the building exists [48]. Other types of infrastructure have been less studied and offer hereto untapped efficiency improvements. Water treatment and distribution, freight transport, the electrical grid, and wireless communications, and others could all benefit from considering engineering performance criteria, agency and user cost, and human factors together for informing improved policy.

REFERENCES

1. Wilson J H, Williams L Z, Schreiber J J, Mullen M A, Peterson T D and Strait R 2009 State approaches to reducing transportation sector greenhouse gas emissions Transportation, Land Use, Planning, and Air Quality: Selected Papers of the Transportation, Land Use, Planning, and Air Quality Conference 2009 (Reston, VA: ASCE) p 9
2. Keoleian G A, Kar K, Manion M M and Bulkley J W 1997 Industrial ecology of the automobile: a life cycle perspective Technical Report (Warrendale, PA: Society of Automotive Engineers (SAE))
3. Hendrickson C T, Lave L B and Matthews H S 2006 Environmental Life Cycle Assessment of Goods and Services: An Input–Output Approach (Washington, DC: Resources for the Future)
4. Thenoux G, Lvaro G and Dowling R 2007 Energy consumption comparison for different asphalt pavements rehabilitation techniques used in Chile Resour. Cons. Recy. 49 325–39

5. Horvath A 2003 Life-cycle environmental and economic assessment of using recycled materials for asphalt pavements Technical Report 510 (Berkely, CA: University of California Press)
6. Hendrickson C, Horvath A, Joshi S, Lave L and Design G 1998 Economic input–output models for environmental life-cycle assessment Environ. Sci. Technol. 32 184
7. International Energy Agency 2009 Transport Energy and CO2 (Paris: IEA Publications)
8. Santero N J and Horvath A 2009 Global warming potential of pavements Environ. Res. Lett. 4 034011
9. Norman J, MacLean H L and Kennedy C A 2006 Comparing high and low residential density: life-cycle analysis of energy use and greenhouse gas emissions J. Urban Plann. Dev. 132 10–21
10. Huang Y, Bird R and Bell M 2009 A comparative study of the emission by road maintenance works and the disrupted traffic using life cycle assessment and micro-simulation Transp. Res. D 14 197–204
11. Golob T F, Recker W W and Alvarez V M 2004 Freeway safety as a function of traffic flow Accident Anal. Prevent. 36 933–46
12. Pawlovich M D, Li W, Carriquiry A and Welch T 2006 Iowa's experience with road diet measures: use of Bayesian approach to assess impacts on crash frequencies and crash rates Transp. Res. Record, J. Transp. Res. Board 1953 163–71
13. CrossRef[13]Ewing R 1993 Transportation service standards—as if people matter Transp. Res. Record, J. Transp. Res. Board 1400 10–7
14. Miller J S and Evans L D 2011 Divergence of potential state-level performance measures to assess transportation and land use coordination J. Transp. Land Use 4 81–103
15. Gu W, Ouyang Y and Madanat S 2012 Joint optimization of pavement maintenance and resurfacing planning Transp. Res. B 46 511–9
16. Lidicker J, Sathaye N, Madanat S and Horvath A 2012 Pavement resurfacing policy for minimization of life-cycle costs and greenhouse gas emissions J. Infrastruct. Syst. at press (doi:10.1061/(ASCE)IS.1943-555X.0000114)
17. Zhang H, Keoleian G A and Lepech M D 2012 Network-level pavement asset management system integrated with life cycle analysis and life cycle optimization J. Infrastruct. Syst. 19 99–107
18. Sathaye N and Madanat S 2012 A bottom-up optimal pavement resurfacing solution approach for large-scale networks Transp. Res. B 46 520–8
19. Fellendorf M 1994 VISSIM: a microscopic simulation tool to evaluate actuated signal control including bus priority 64th Institute of Transportation Engineers Annual Mtg
20. Faghri A and Egyhaziova E 1999 Development of a computer simulation model of mixed motor vehicle and bicycle traffic on an urban road network Transp. Res. Record, J. Transport Res. Board 1674 86–93
21. Behrisch M, Bieker L, Erdmann J and Krajzewicz D 2011 SUMO—simulation of urban mobility: an overview SIMUL 2011: 3rd International Conf. on Advances in System Simulation (Barcelona, Oct. 2011) pp 63–8

22. Noland R B 2003 Traffic fatalities and injuries: the effect of changes in infrastructure and other trends Accident, Anal. Prev. 35 599–611
23. Johan de Hartog J, Boogaard H, Nijland H and Hoek G 2010 Do the health benefits of cycling outweigh the risks? Environ. Health Perspect. 118 1109–16
24. Persaud B and Lyon C 2010 Evaluation of lane reduction 'Road Diet' measures on crashes and injuries Technical Report 1 (Washington, DC: Federal Highway Administration)
25. Dowling R, Flannery A, Ryus P, Petrisch T and Rouphail N 2010 Field test results of the multimodal level of service analysis for urban streets National Cooperative Highway Research Program Web-Only Document 158 (Washington, DC: Transportation Research Board)
26. American Automobile Association 2007 Fuel Gauge Report
27. AASHTO 2010 User and Non-User Benefit Analysis for Highways (Washington, DC: American Association of State Highway and Transportation Officials)
28. Transportation Research Board 2010 Highway Capacity Manual (Washington, DC: Transportation Research Board of the National Academies)
29. Rosey F, Auberlet J-M, Moisan O and Dupre G 2009 Impact of narrower lane width: comparison between fixed-base simulator and real data Transp. Res. Record, J. Transp. Res. Board 2138 112–9
30. Zorn W 2008 Speed and Power Calculator (www.kreuzotter.de/english/espeed.htm, referenced 20 October 2012)
31. Barnes G and Langworthy P 2003 The per-mile costs of operating automobiles and trucks Technical Report (St Paul, MN: Minnesota Department of Transportation)
32. Zaabar I and Chatti K 2010 Calibration of HDM-4 models for estimating the effect of pavement roughness on fuel consumption for US conditions Transp. Res. Record, J. Transp. Res. Board 3 105–16
33. Karlsson R, Hammarström U, Sörensen H and Eriksson O 2011 Road surface influence on rolling resistance Coastdown measurements for a car and an HGV Technical Report
34. Barth M, An F, Younglove T, Scora G, Levine C, Ross M and Wenzel T 2000 NCHRP project 25–11 development of a comprehensive modal emissions model National Cooperative Highway Research Program Web-Only Document 122 (Washington, DC: Transportation Research Board)
35. Gosse C, Smith B and Clarens A 2012 Environmentally preferable pavement management systems J. Infrastruct. Syst. at press (doi:10.1061/(ASCE)IS.1943-555X.0000118)
36. Fekpe E 2006 Payment damage from transit buses and motor coaches Technical Report (Columbus, OH: Batelle Memorial Instititue)
37. Diefenderfer B K, Moruza A K, Brown M C, Roosevelt D S, Andersen E C, O'Leary A A and Gomez J P 2010 Development of a weight-distance permit fee methodology for overweight trucks in Virginia Int. J. Pavement Res. Technol. 2 236–41
38. Shoup D 2011 The High Cost of Free Parking 2nd edn (Washington, DC: American Planning Association)

39. Jacobsen P L, Racioppi F and Rutter H 2009 Who owns the roads? How motorised traffic discourages walking and bicycling Injury Prev.: J. Int. Soc. Child Adolescent Injury Prev. 15 369–73
40. 2006 Bicyclist fatalities and serious injuries in New York City 1996–2005 Technical Report (New York: New York City Departments of Health and Mental Hygiene, Parks and Recreation, Transportation, and the New York City Police Department)
41. Works Department and Emergency Services 2003 City of Toronto bicycle/motor-vehicle collision study Technical Report (Toronto: Transportation Services Division Transportation Infrastructure Management Section)
42. Noland R B and Lem L L 2002 A review of the evidence for induced travel and changes in transportation and environmental policy in the US and the UK Transp. Res. D 7 1–26
43. Rahul T M and Verma A 2013 Economic impact of non-motorized transportation in Indian cities Res. Transp. Econ. 38 22–34
44. Lemer A C 1996 Infrastructure obsolescence and design service life J. Infrastruct. Syst. 2 153–61
45. 2011 NYC commuter cycling indicator Technical Report (New York: New York City Department of Transportation)
46. 2005 District of Columbia bicycle master plan Technical Report (Washington, DC: District Department of Transportation)
47. Krizec K 2007 Estimating the economic benefits of bicycling and bicycle facilities: an interpretive review and proposed methods Essays on Transport Economics (Contributions to Economics) ed P Coto-Millán and V Inglada (Heidelberg: Physica-Verlag) pp 219–48
48. US Green Building Council 2009 LEED Reference Guide for Green Neighborhood Development LEED 2009 edn (Washington, DC: US Green Building Council)

PART IV

URBAN WATER SUPPLY

CHAPTER 10

Discussion on Sustainable Water Technologies for Peri-Urban Areas of Mexico City: Balancing Urbanization and Environmental Conservation

TIEMEN A. NANNINGA, IEMKE BISSCHOPS, EDUARDO LÓPEZ, JOSÉ LUIS MARTÍNEZ-RUIZ, DANIEL MURILLO, LAURA ESSL, AND MARKUS STARKL

10.1 INTRODUCTION

In urban and densely populated areas, centralized solutions for water management are usually favored over small scale solutions based on principles of economy of scale and economy of density [1,2,3]. In this traditional approach, water, wastewater and solid waste are often transported over large distances. These centralized water supply, sanitation and solid waste solutions often struggle to keep up with rapid expansion of large cities in the South, such as Mexico City, not only due to unregulated and uncontrolled urbanization, but also due to high construction costs [4]. Indeed, it is in these peri-urban areas that there is a trend towards providing water supply, sanitation and solid waste management technologies on a decentralized, on-site scale [4,5,6,7,8]. However, stakeholders may prefer centralized solutions because of the convenience of a centralized system.

The introduction of decentralized technologies allows for the development of new opportunities that enable the recovery and reuse of resources in the form of water, nutrients and energy. This resource-oriented management of water, nutrients and energy requires a sustainable system aimed at low resource use and high recovery and reuse rates.

Instead of designing each sector separately, which has been traditionally done, this article proposes and discusses a concept that combines the in- and outflows of the different sectors, reusing water and other liberated resources where possible, which is illustrated in Figure 1. "Joints" for possible integration of these sectors are: water reuse, nutrient recycling and energy recovery. In the decentralized concept all wastewater in a region is seen as a separate water, nutrient and energy source, and is evaluated for its suitability as a water source for a specific use, such as agriculture, non-potable domestic purposes or forest irrigation. With letting the final users determine the quantitative and qualitative water requirements needed for a specific purpose such as agriculture, industry or (non-potable) household activities, the user or purpose of the wastewater determines the quality to which the wastewater is improved, which is the opposite of the conventional water chain where governmental legislation or permits dictate the treatment level. This reverse water chain [9,10,11], results in a new and novel view of waste and wastewater treatment and water sourcing where legislation should allow for tailor-made solutions. This may also reduce costs as not all waste and wastewater need to be treated up to high standards, since a part can also be used for purposes only requiring lower standards.

This paper shows and demonstrates examples of different types of sustainable water technologies that can be implemented in the peri-urban areas of Mexico City. An innovative participatory planning method using a scenario building methodology at project level was applied. This method has allowed for a meaningful and intensive involvement of a variety of key stakeholders in the planning process. It also helped to identify the opportunities and limitations of resource oriented water technologies within an urban planning context highlighting the conflict between environmental conservation and urbanization.

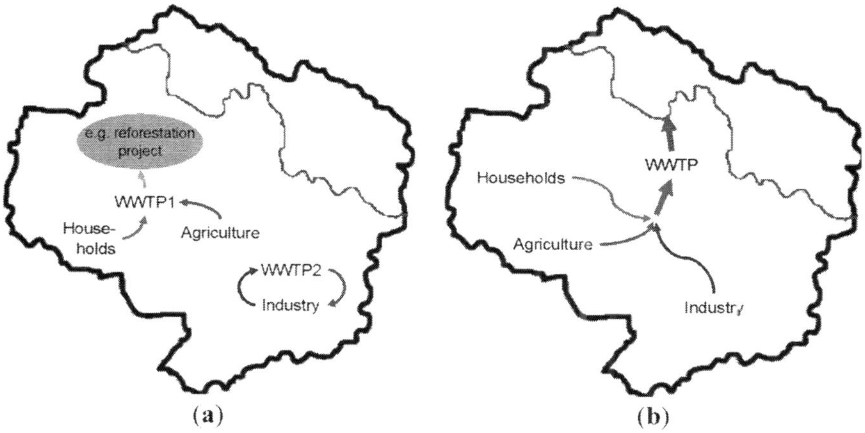

Figure 1. (a) Traditional way of thinking about wastewater treatment; and (b) resource oriented way of thinking on a regional scale. WWTP1 and WWTP2 can constitute completely different technologies (WWTP = Wastewater Treatment Plant).

10.2 CASE STUDY

The case study is located in Xochimilco, one of the 16 boroughs (delegaciones) in Mexico City. The entire delegación Xochimilco has an area of approximately 12.5×10^7 m^2 [12,13], of which 2.5×10^7 m^2 are urbanized. The peri-urban areas (which can be legal or illegally urbanized) cover the additional 3.0×10^7 m^2, and serve for this study (Figure 2). These peri-urban areas have a mixed topography with plain (1.26×10^7 m^2) as well as mountainous areas (1.74×10^7 m^2). Xochimilco has a high biological and cultural value due to the wetlands and chinampas. Chinampas are rectangular land plots (usually 10 by 100 meters) surrounded by narrow canals used for irrigation, fishing, transportation and as a water source. The construction of chinampas dates from pre-colonial times and they are used for agricultural purposes [12,13,14,15]. In general, they are considered to be biologically very diverse and agriculturally very high productive areas [12,16,17]. Farmers cultivating crops on a chinampa are called

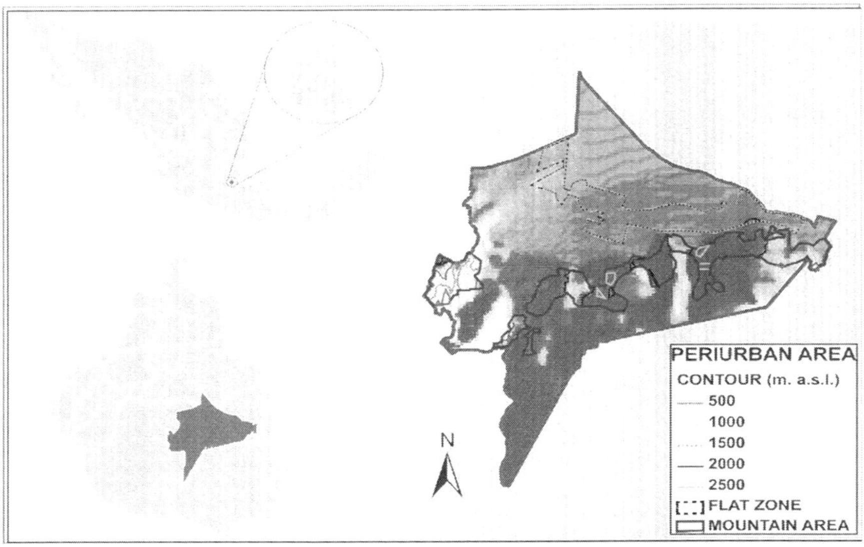

Figure 2. Case study area of Mexico City, consisting of the peri-urban zones of Xochimilco.

chinamperos. It was listed as a World Heritage site in 1986 and as a wetland of international importance in 2004 [12,18].

The trends of the past four decades confirm a dual-pressure process in Xochimilco, referring to progressive shortages of water and expansion of other land uses. Natural systems (water interactions, soil, vegetation, fauna) are under pressure by the urbanization. In recent years, water has been injected into the aquifer and is used as the main water supply source to retard groundwater depletion of Mexico City. The effluent of a wastewater treatment plant, Cerro de la Estrella, is discharged to the surface water bodies of Xochimilco to ensure that water remains [12]; if this is not done, the surface water bodies will dry as the natural springs have been depleted [19]. Nevertheless, there are no signs that the drying effect on both groundwater and surface water can be countered, not to mention the environmental pollution that takes place due to the discharge of untreated wastewater by households living near or on the chinampas. If this trend continues, the expected scenario for the area is that there will be a large water deficit in the next century [20]. The peri-urban zone has been experiencing problems with services for water, wastewater and solid waste management. This situation will be aggravated due to the unregulated urbanization, definitely.

10.3 METHODOLOGY

10.3.1 PARTICIPATORY PLANNING

Participatory planning is considered an important aspect for achieving sustainable water services, e.g., [21,22,23]. In this project an innovative approach using scenario building methodology was applied. From the wide range of available methods for scenario building, in this project we were in particular interested in those which allow the users to participate in shaping the development of their region. An example for such a method is the Future Workshop (FW) method. This method allows participants to become involved in creating their preferred future. A "classic" FW, according to Jungk and Müller [24], consists of five phases:

(1) The preparation phase;
(2) The critique phase;
(3) The fantasy phase;
(4) The implementation phase;
(5) The follow-up phase.

In this study, the scenario workshops conducted under this study encompassed the first three phases, but can be adapted to the local needs and situation as follows:

(1) Presentation of the overall study;
(2) Presentation of urban problems in the case study area;
(3) Presentation of trend analysis study;

(4) Participants wrote what they like and what they do not like about the current situation on cards;

(5) Elaboration of vision "Xochimilco 2030" in three mixed working groups;

(6) Presentation of visions.

The scenario workshop aimed at the identification of different options for future regional development. Building on the outcomes of the scenario workshops, a workshop for participatory planning was conducted. This workshop focused more on the technical aspects with respect to water, wastewater and solid waste management. It encompassed two main phases: the existing environmental problems in the area were discussed; and the participants identified possible solutions and highlighted the main conflicts and barriers that need to be overcome to implement those solutions.

10.3.2 DEVELOPMENT OF CONCEPT SCENARIOS

The results of the scenario building and the participatory planning were combined and then different concept scenarios were developed. A concept scenario encompassed a coherent set of water technologies that are suitable for a different future development scenario (e.g., urbanization or conservation). The concept scenarios were then furnished with a set of suitable technologies.

10.3.3 TECHNICAL FEASIBILITY STUDY

Next, a feasibility was conducted which aimed at demonstrating the technical feasibility of the identified technologies for each concept scenario. As a detailed feasibility study for the entire case study area was beyond the scope of this study, a smaller area was considered much better to suit for "testing" the concepts and its technologies. For the selection, some criteria including infrastructure, urbanization, remoteness and socio-economic conditions were applied to ensure the selection is representative for most peri-urban area in Xochimilco. This included one small area in the lowland chinampa area (La Conchita) and another in the mountainous areas (San Martín). The detailed feasibility study was then conducted for each concept scenario in the selected smaller areas. The following tasks were conducted:

1. A detailed survey of the existing infrastructure in the case study area and a household level survey in the selected smaller areas.
2. A detailed technical feasibility study, which included technical design and drawings of the set of technologies within each concept scenario, thus demonstrating their technical feasibility.

10.3.4 RESOURCES FLOW ANALYSIS

The flows of the different resources were analyzed qualitatively and mapped out for each concept scenario. The resources analyzed were water, Nitrogen (N), Phosphorus (P), organic matter (OM) and energy. Then for selected technologies calculations were performed to quantify the flows. The input data used for these calculations are shown in the Table 1, Table 2 and Table 3 below:

The analysis and calculations not only show how (local) finite resources such as water, P and energy are recovered from waste (water), but also how they can be reused. N and OM give a clear indication of how pollution of surface water bodies can be lessened and, with P, how valuable nutrients can be recovered and reused. The analysis of the resources was grouped in water, nutrients and energy, and focused on the import and export resources as well as the generation, recovery and usage of resources within the small case study areas.

10.4 RESULTS

10.4.1 RESULTS—PARTICIPATORY PLANNING

With the past and current systems in Xochimilco, natural resources have been using wastefully, continuously. The case of Xochimilco is critical for the future of whole Mexico City as it serves as a strategic water reserve for the city and for aquifer recharge. As shown in the introduction there

Table 1. Amounts and composition of urine and feces [25].

Waste stream	Amount	N (g/L)	P (g/L)	K (g/L)
Urine	1.2 l/p.d	3	0.8	1.3
Feces	150 g/p.d	2	0.6	0.6

Table 2. Amount of organic solid waste and general characteristics of compost [26,27,28].

Waste stream	Amount	Density	N (g/L)	P (g/L)	OM (g/L)
Organic solid waste produced in Xochimilco	650 g/p.d	-	n/a	n/a	n/a
Compost	-	450 kg/m^3	3.2	0.6	120*

* Calculated from [27,28].

Table 3. Precipitation data (mm) for San Gregorio Atlapulco weather station, Xochimilco [29].

Precipitation	Jan.	Feb.	Mar.	Apr.	May	Jun.	Jul.	Aug.	Sep.	Oct.	Nov.	Dec.
Monthly sum*	13.4	14	11.7	30.7	78.6	147.9	172.2	148.7	111.4	64.2	7.8	6.6
Daily maximum	32.0	20.3	17.8	23.9	36.8	46.8	49	63.4	36.5	42.0	17.5	14.5

* Monthly sum are according to climatological normals (1971–2000) and daily maximum are observed.

is a strong conflict between urbanization of Xochimilco and environmental conservation, a scenario workshop was conducted regarding this background. The participants were from different stakeholder groups including chinamperos, local government representatives, NGOs and academics. The visions of the participants were similar, but could not be summarized in one single concept. Conservation of identity, which was very often mentioned during the workshop, can be partly reached by preserving water resources as they are very important part of the local culture of the chinamperos. Economy, society and the environment should be the vertices of a triangle, and have equal value. In the center of that figure may be awareness, culture and education. They could be starting points for technologies that consider also the tradition of the area and constitute an identity and culture focusing on social justice and a balanced environment applying criteria of sustainability. The engine of these processes should be presence of and pressure from citizens: promoting the creation of new spaces and forms of citizenship or re-elaborate some of the dynamics forms of identity and space. Community-based initiatives can be more effective than public policies supported by governmental institutions, unable to sustain over several administrations.

A synthesis of the visions has shown two possible development scenarios (a development scenario is a vision of the users and stakeholders that participated in the Future Workshop on the future development of the area):

1. Conservation of local identity: as this aspect was mentioned very often during the workshop, a scenario which strengthens the local community was developed. It envisages local solutions and some level of engagement with the city government where inhabitants of Xochimilco are involved and participating in decision making processes concerning resource management. Water protection is a crucial component as water is highly connected to the local identity and the traditional form of cultivation in chinampas.
2. Economic development: This scenario builds on economic development. Originally the focus was on touristic development; however, a later refinement to the case study areas has put emphasis on agricultural development. Although the productivity of the chinampas is generally high [12], not all are used for agricultural purposes. Instead, some parts are fallow and others are used for urbanization. By protecting the chinampas from urbanization and promoting agricultural activities through capacity building and for instance new high-value crops income can increase. In addition, Xochimilco has a touristic area that could be promoted and improved. Prerequisites are a functioning water supply and sewage net, a solid waste management system and the protection of biodiversity and the natural environment that makes the area for tourists an attractive, clean destination. Local products could be certified and sold in supermarkets in Mexico City to generate higher income.

Then, a planning workshop that focused on the technical aspects related to water, wastewater and solid waste management was conducted. During the workshop participants defined various problems related to the water supply, wastewater, agriculture and solid waste for peri-urban areas in Xochimilco that they experience or perceive.

Participants acknowledge the water supply problems, including depletion of the aquifer and poor water supply services. In addition, they also indicate that the water quality, both of the environmental water as well as the domestic water, is poor, which induces several health problems. The participants perceive the discharge of poorly or un-treated wastewater to the canals, streets and environment as the main problem. This results in water pollution, loss of biodiversity and health problems. The performance of the wastewater treatment plants is perceived as being poor, both in terms of quantity and quality. The main agricultural problems are related to a lack of knowledge of farmers and the loss of agricultural land and indigenous agricultural practices such as the chinampas. The absence of solid waste classification, collection and recycling is perceived as a problem, which is mainly related to the lack of awareness about the possibilities for separating and recycling solid waste. Participants have formulated possible technical solutions to the above mentioned problems. They proposed to introduce alternative technologies for water supply, such as rainwater harvesting and groundwater recharge systems. Filters for grey water and the installation of dry toilets could improve the situation concerning untreated wastewater discharge. For solid waste management, the initiation of a waste separation program was proposed by the participants.

10.4.2 RESULTS—CONCEPT SCENARIOS

Each concept scenario is related to one of the possible development scenarios and encompasses a set of technologies that is suitable for the development scenario. In addition, a new concept scenario has been created that is related to increased urbanization. Even if that was not preferred by the participants of the scenario workshop it is a very realistic scenario as a recent trend analysis based on GIS data has shown (see Figure 3). The identified three concept scenarios are summarized in Table 4.

Table 4. Key objectives and characteristics of the three concept scenarios.

Scenario	Key objectives	Characteristics of technologies under this scenario
1. Local identity	The goal of this scenario is the conservation of local identity which is related to the cultivation of chinampas and the prevention of external influences.	In this concept scenario individual technical solution are preferred over centralized ones to become more independent from Mexico City.
2. Economic development	The goal of this scenario is economic development with a strong focus on agriculture.In the mountainous areas where no agriculture is practiced, there is a focus on community development.	In this scenario there is a strong emphasis on sanitation systems that allow the reuse of nutrients and the water in the chinampas or in some other areas to improve the agricultural production.In the hilly area, community technologies are the main feature of this scenario.
3. Centralization	The main goal is a strong connection to the development of Mexico City and integration into the planned urbanization.	All infrastructure services are centralized as much as possible.

Table 5. Examples for technologies with a focus on reuse and recycling.

Resource	Proposed treatment technology
Water	Rain Water Harvesting;
	Constructed wetlands;
	Compact treatment plants (e.g., Biostar);
	(Biological) filters to treat grey wastewater or canal water prior to use.
Energy	Anaerobic digestion of organic waste
Nutrients	EcoSan systems;
	Composting;
	Vermicomposting.

The proposed technologies for concept scenarios 1 and 2 have a strong focus on the reuse and recycling of water, nutrients and energy. Table 5 shows examples for technologies that allow a recovery of nutrients, energy and reuse of water.

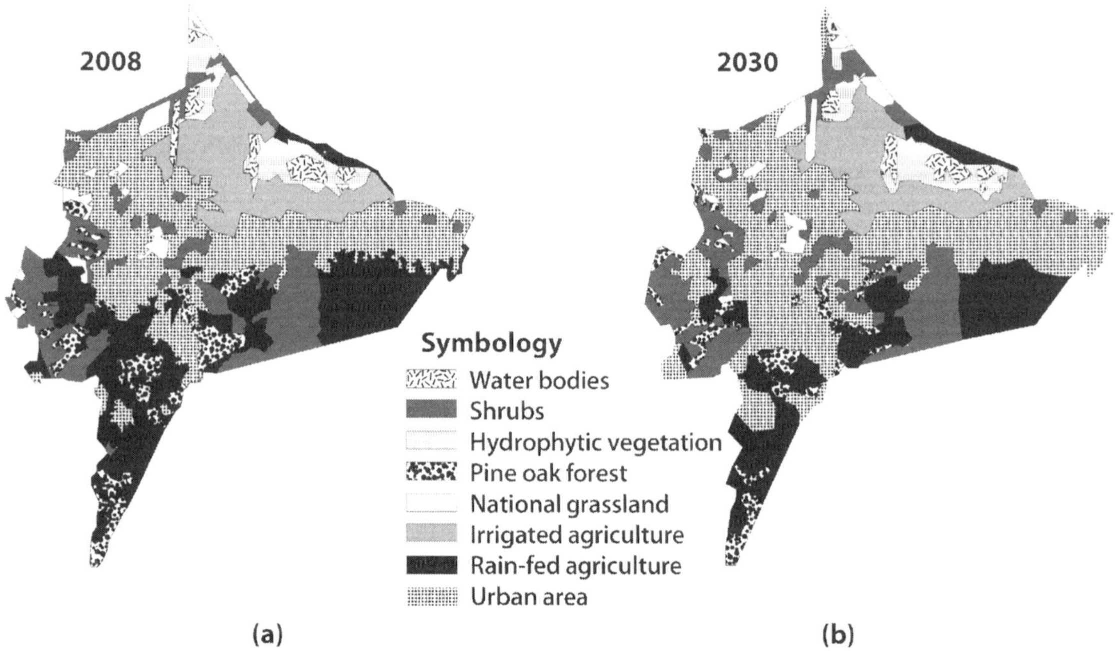

Figure 3. (a) Map of Xochimilco 2008; and (b) trend analysis for 2030 (Adapted from [30]).

10.4.3 RESULTS—TECHNICAL FEASIBILITY STUDY

10.4.3.1 WATER SUPPLY

Currently, as rainwater is the only water source not yet exploited in Xochimilco, special attention was paid to the technologies focused on this. The harvesting, storage and treatment (UV treatment, chlorination) of rainwater to supply water for households during the raining season, in combination with gabion dams that are constructed at strategic locations and thereby facilitate the infiltration of collected surface runoff, are two technologies that make optimal local use of the rainwater. The use of rainwater in the households lessens the demand for water from other sources (i.e., the aquifer, surface water bodies) during a certain time of the year, and the infiltration of the remaining rainwater enables the replenishment of the aquifer and hence storage of the water for times when the water demand is higher than the available precipitation. The use of treated wastewater for non-potable domestic and agricultural purposes also lessens the demand for potable water. In general little polluted wastewaters, such as grey wastewater, can be treated by decentralized wastewater treatment systems before reuse without posing any environmental or health risks. Water from canals can also be used for specific non-potable domestic purposes such as cleaning or watering of ornamental plants. The canal water in Xochimilco is too polluted to be used directly for domestic purposes, but can be treated by for instance a constructed wetland or small water filter depending on its quality and the proposed use. Because the main precipitation in Xochimilco falls in six months (May till October), the existing centralized water supply system is expected to remain a very important source for drinking water. The conventional water supply system uses groundwater as its source, which indicates the importance of groundwater replenishment technologies. The rehabilitation and expansion of the water supply system by, for instance, legalizing the illegal connections to the water supply lines and ensuring that the existing (illegal) water pipes are properly constructed and maintained and that taps are not left to run unattended, hence remains an important action. The following example shows how the demand for water from the centralized water supply system can be lessened through the implementation of on-site rainwater harvesting technologies with UV-treatment prior to use or consumption.

10.4.3.1.1 DESCRIPTION OF CURRENT SITUATION

Although workshop participants have, amongst others, proposed the use of rainwater harvesting technologies (RWH) (see Section 4.1 and Table 5), at the moment these are not applied in the case study area. This can be attributed to the lack of capacity needed to construct and operate such a system, unwillingness to implement a new technology and/or the poor state of several houses. Houses in the peri-urban areas are built with a variety of materials, including cement, ceramic tiles and corrugated iron sheets, asbestos sheets, cardboard and wood. In general, inhabitants of the peri-urban areas do not pre-treat their water before consuming unless boiling. The application of tUVo's (a UV-treatment technology developed in México) and chlorination of water prior to use is currently not common.

10.4.3.1.2 POTENTIAL APPLICATION IN CASE STUDY AREA

The harvesting, storage and utilization of rainwater at domestic level is an alternative to avoid the overexploitation of the underground aquifers and the surface water sources in the peri-urban areas of the México City. This will be possible in the raining season and part of the dry season (December). The average annual precipitation in Mexico City is 807 mm [29], with the majority falling in June till October. The collected rainwater will be stored in storage tanks before use. The capacity of these storage tanks will depend on the water demand as well as if there are also connections to the centralized water supply system. As the average weekly household water demand is 0.8 m^3 (with each household consisting of 4 individuals and based on households with flush toilets as well as pit latrines) a total water amount of 3.5 m^3 is needed per month. In general, the roof surfaces of houses in the peri-urban areas of Xochimilco are estimated to be around 36 m^2 [31]. Assuming a 70% collection efficiency (losses and diversion of the first flush) 4.3 m^3 of rainwater can be harvested in the month with the highest monthly precipitation (July; 172 mm) and 0.2 m^3 in the month with the lowest precipitation (December; 7 mm). The highest monthly RW supply is almost equal to the monthly water demand, which means that the storage tank can be kept small as water will be used the same month it is collected. The highest daily maximum precipitation is around 63 mm, yielding a collected amount of 1.6 m^3. By installing a storage tank with the minimum 1.5 m^3 for each household, it is ensured that all collectable rainwater can be utilized optimally.

The installation of RWH technologies can result in the improvement of the living conditions of the inhabitants due to the improved roofs and supporting structures indirectly, and water saving directly. This will also affect the environment indirectly, as less water will be extracted from the aquifer or canals.

The installation of on-site RWH technologies will require the adaptation or improvement of houses. The main adaptations required will be new roofs and supporting structures if these are not suitable for rain water harvesting. In addition, gutters, transportation pipes and a storage tank will need to be installed. Installing the storage tank on a structure (but below the level of the roof) the water can be collected with pales or transported into the house with a pipeline. The water can be treated in the house with filtration, and/or a UV-light (a tUVo, requiring energy) or chlorination.

10.4.3.2 WASTEWATER TREATMENT

The use of EcoSan toilets could improve the sanitation conditions of the households and do not require water, as is the case with flush toilets, thereby lessening the overall household water demand. This is an advantage due to the current aquifer depletion taking place. Lastly, the advantage of EcoSan is that the feces and urine can be co-composted with other organic wastes or the effluents can be applied directly on the fields (chinampas or greenhouses), thereby containing resources in the area. Discharged wastewater that is not very polluted, such as grey wastewater, can be treated by constructed wetlands that can be constructed in the yards of the houses.

Another possibility is to collect the wastewater and treat it with a decentralized wastewater treatment system such as a constructed wetland.

These can be constructed in gullies or drains where wastewater is collected or discharged to. Another option is to implement a more complex technology such as a Biostar, which is a biological wastewater treatment, designed by the Instituto Mexicano de Tecnología del Agua (IMTA) in Cuernavaca, México, or an Rotating Biological Contactor (RBC). However, these will need a (small) sewage network and perhaps pumps to collect and transport the wastewater besides the need for skilled operators as well as an energy source to function and perform adequately.

In a more centralized scenario the existing sewage system and wastewater treatment plant can be extended. However, the multiple unplanned settlements not only make it difficult to extend the existing sewage system in order to serve all households (as the capacity of the wastewater treatment plants might need to be increased and a new sewage network will need to be constructed amidst existing buildings and chinampas, thereby severely limiting the design possibilities), but there are also multiple buildings constructed over the existing sewage system which makes rehabilitation and extension works impossible to perform without demolishing and reconstructing the buildings. Hence the expansion of the centralized system might be more difficult to realize than the construction of decentralized measures. On the other hand, the centralized system is a technology preferred by the local inhabitants, meaning that the implementation of decentralized sanitation options might be more difficult due to the lower social acceptance. This, however, still needs to be evaluated. An example of how the implementation of a decentralized sanitation option such as EcoSan can result in resource recovery and water conservation is presented as follows:

10.4.3.2.1 DESCRIPTION OF CURRENT SITUATION

The sanitation situation in the peri-urban areas of Xochimilco differs per location. A number of houses, generally located near the urban areas and that are connected to the water supply system or store water in tanks, make use of flush toilets. These discharge their water into the canals or septic tanks, which are often poorly constructed or just a hole in the ground. Other, more remote, households make use of pour flush toilets or pit latrines that are often poorly constructed or defecate and urinate directly in the environment. Gastronomical and skin problems and diarrheal diseases are common in these areas [32].

10.4.3.2.2 POTENTIAL APPLICATION IN CASE STUDY AREA

By constructing EcoSan toilets initially at households that currently do not have sanitation facilities or make use of pit latrines access to proper sanitation facilities is improved. The combination with a biofiltro, which is the local name for a small constructed wetland designed to treat grey wastewater at household level, will ensure that all wastewater originating in a household is adequately treated. As it is generally the poor who do not have proper access, their livelihoods and the overall public health is impacted directly and improved. In addition, the recovery and reuse of nutrients can not only lead to less demand for artificial fertilizer (or if no fertilizer was used, higher yields), but it can also improve the soil conditions and thereby the sustainability of the land.

An EcoSan toilet can be constructed in the yard of a household, or as an extension of a house, where in general the dimensions of a toilet or pit latrine (1–1.5 m^2) can be maintained. Care should be taken to design the EcoSan toilet in such a way that it uses energy from the sun to dry the collected feces. Feces and urine will be stored in two separate containers prior to their use or the co-composting. The feces and urine can be used in local gardens, greenhouses or chinampas in the form of compost. A biofiltro, with a surface area of 2 m^2 to treat 100 L of grey water (total dissolved oxygen demand of 555 mg/L) per day [11], can also be constructed in the yard of a house. In certain areas this is already common, and here they are constructed in the form of flowerbeds. Cases where ornamental flowers (Zantedeschia aethiopica and Cannaflaccida) are used, with good treatment results, are also known [33].

10.4.3.3 SOLID WASTE MANAGEMENT AND AGRICULTURE

Organic waste management and agriculture are treated in one section as the outputs of the organic waste treatment often can be used in agriculture.

Waste separation practices are considered a key practice for solid waste management. By separating the different wastes, some can be reused or recycled, while others can be used as an input for composting in order to retain nutrients in the area. If the waste is collected by a centralized organization, the inhabitants will need to store their wastes in separate closable bins till the wastes are collected. The wastes can then be recycled, composted or disposed of in a location specifically adapted to the purpose. For instance, a centralized composting plant, using vermicomposting, could treat organic wastes from an area and in turn supply the farmers of that area with a high-quality compost, soil improver or organic fertilizer. Depending on the local demand for energy, it is also possible to digest the organic wastes with manure coming from livestock to produce biogas. This can be done on household level or in a centralized way. The effluent of the biodigester can be used for fertigation and composting purposes, or organic fertilizer can be produced.

On-site composting and vermicomposting can be implemented where there is no, or very poor, waste collection facilities as the produced compost can be handled safely. These practices can also be done by families with gardens so that they can reduce their fertilizer demands. Although vermicomposting is more efficient and can eventually produce valuable side-products such as worms and proteins, it also requires more economical input (worms—i.e., Eisenia fetida—need to be bought, a structure needs to be constructed to contain the compost) to start-up, and thus by studying both technologies local inhabitants can choose which composting type they prefer and can afford. The location of the (vermi-)composting can be in the vicinity of the house, to ease organic waste disposal, but at a large enough distance to prevent the occurrence of noxious odors or pests. The following is the presentation and discussion of vermicomposting as an example of on-site organic waste management:

10.4.3.3.1 DESCRIPTION OF CURRENT SITUATION

The separation and recycling of solid waste is an uncommon practice in the peri-urban areas of Xochimilco. Although a number of inhabitants have access to the centralized collection service, who collect the solid waste with trucks, there are also households who do not make use of these facilities, are not or irregularly visited by the trucks or simply do not make use of the service. Hence solid waste is also burnt or discarded to the streets and environment, resulting in pollution.

10.4.3.3.1 POTENTIAL APPLICATION IN CASE STUDY AREA

The local separation of solid waste and local vermicomposting of organic solid waste will result in an overall decrease of the amount of waste to be collected. This, in turn, will lower the strain on the municipal waste collection service so that they can extend their services. Another advantage is that awareness about waste production and treatment is created with the inhabitants, making them more aware of their actions and hence less inclined to pollute the environment. Lastly, valuable nutrients are conserved and can be used in the chinampas and gardens as natural fertilizer and soil conditioners.

Vermicomposting is technologically relatively simple to conduct. It will, however, require a behavior change as solid waste will need to be separated. Also, the organic solid waste will be composted near the house, in the yard, which is something that not all households are eager to do, especially as the compost pile might need to be over-turned for active aeration. If it is done, one will need to construct a box (usually 1 m3). The composting pile will need to have a relative humidity of 70%, and an average temperature of 21 degrees Celsius for the reproduction of the worms. However, during the composting process temperatures can exceed 60 degrees Celsius, which will result in a reduction of pathogens. During (vermi-)composting in general a mass reduction of 10%–30% can be expected [34,35].

10.4.4 RESULTS—RESOURCES FLOW ANALYSIS

Based on the three concept scenarios identified during the workshop and the technologies allocated to each concept scenario resources flows were identified and calculated.

The goal of the concept scenario "Local Identity" is the conservation of local values which is related to the cultivation of chinampas and the prevention of external influences. In this concept scenario individual technical solution are preferred over centralized ones to become more independent from Mexico City.

By implementing multiple on-site technologies waste streams can be collected and treated separately (Table 6). In this concept scenario various resources flows between the sectors will take place: nutrients in the form of urine and decomposed feces as well as treated wastewater and compost will be used in agriculture. The digestion of organic waste produces biogas which is an additional local energy source. This will not only enable the recovery and reuse of water, nutrients and energy, but also lessen the demand for centralized services and hence create more independence for the users.

The objective of the concept scenario "Economic and community development" is economic development with a strong focus on agriculture. In the mountainous areas where no agriculture is practiced, there is a focus on community development. In this scenario (Table 7) the main resources flows emanate from the sanitation systems that allow the reuse of treated wastewater in the chinampas or in some other areas to improve the agricultural production and to have better tourist services. The compost of the decentralized composting facility is used in local agriculture. In the mountainous area, community technologies are the main feature of this scenario.

The advantage of implementing technologies on communal scale is that larger quantities of water and nutrients can be recovered at one location. Hence it is possible to reuse these resources as inputs for agriculture. In addition, side benefits, such as a business in composting, can be established. However, it is more of a challenge to separate all waste streams from the source as certain wastes have specific challenges. Urine, for example, does not allow for transportation over long distances through pipes as it can result in clogging due to precipitation of minerals (primarily Calcium and Magnesium Phosphates) [36].

In the concept scenario "Centralization" the main goal is a strong connection to the development of Mexico City and integration into the planned urbanization. All infrastructure services are centralized as much as possible.

In a centralized scenario it is much more difficult to recover resources from wastewater and solid waste (as shown in Table 8). The only resources flow between the sectors is through the organic fraction, which can be composted in a centralized location before being solid back to the farmers. The treated wastewater will be discharged to the environment.

Examples of resources flows from earlier discussed technologies are as follows: Through the implementation of an on-site RWH system with

Table 6. Overview of technical systems for concept scenario1—Concept of natural resources flows.

Concept scenario 1—Local identity			
Water supply	Wastewater	Agriculture	Solid waste
- RWH (on-site and groundwater recharge - Aquifer recharge through rainwater capture - Reuse of treated wastewater for non-drinking purposes	- Separation black water/ grey water - Suitable treatment for black waters (on-site treatment) - Treatment of grey water for reuse	- Conservation of chinampas - Use of compost - Local gardens (to support subsistence with vegetables) - Use of treated waste-water	- Separation organic and inorganic waste - On-site (vermi) composting - Anaerobic diges-tion of organic waste

Table 7. Overview of technical systems for concept scenario 2—Concept of natural resources flows.

Concept scenario 2—Economic and community development			
Water supply	Wastewater	Agriculture	Solid waste
- Communal RWH - Use of treated water from channels for non-drinking purposes ==>	- Community scale wastewater treatment plant (reuse in agricul-ture or for non-domestic purposes)	- Reuse of water and nutrients for chinam-pas for agricultural production - Manure and other residues to composting	- Separation of organic waste - Decentralized composting together with sediments from chinampas and other agricultural residues - Collection of inorganic waste, recycling and disposal

post-treatment the inhabitants will be less depended on the centralized water supply system for their water supply. If all annual precipitation (807 mm) is collected with a 70% efficiency on the before mentioned roof surface (36 m2), this can result in a capturing of 20.3 m3, which accounts for 49% of the annual water demand of a peri-urban household not connected to the centralized water supply system (41.6 m^3/year) [32].

The implementation of an EcoSan and biofiltros for the treatment of feces, urine and grey water at household level will enable the recovery of N, P and Potassium (K) as well as water. Based on data for one household the recovered amounts can be 5.7 kg N/year, 1.5 kg P/year and 2.4 kg K/year. Based on N-requirements, this is enough to cultivate 750 m2 of hybrid maize capable producing 4.5 t/ha (application load of 80 kg/ha). However, if P-availability is limiting, the cultivatable area is 500 m^2. For K this is 300 m^2 [37,38]. This can deviate, depending on local conditions.

If households that currently make use of flush toilets (using an average of 7 L per flush) also construct EcoSan toilets, water can be saved. With an average of 4 inhabitants per household and 5 flushes per person per day (one flush per visit), an annual 51 m^3 can be saved per household. The use of a biofiltro will ensure that little polluted wastewater can be reused in a safe way for, for instance, irrigation of crops or non-potable domestic purposes such as laundry or cleaning. The annual production of grey wastewater for a household is 112 L per household per day, which will result in an annual amount of 31 m^3 of treated wastewater, assuming a 75% recovery and treatment efficiency.

The annual amount of organic waste produced per household in Xochimilco is approximately 1000 Kg. Assuming a 20% mass reduction this results in an annual production of 800 kg of compost from the organic waste of one household. The composition of compost generated will differ per location as it depends upon the organic waste that is composted, the quality of the composting facility (aeration, temperature regulation) as how the compost is maintained (moisture content and temperature). Taking the general values shown in Table 2 the amount of nutrients recovered annually by one household is 6 kg N and 1 kg P. Compost is also a source for organic matter (OM), which can be used as soil conditioner. The annual production of OM is calculated to be 200 kg OM. Based on N-requirements, this is enough to cultivate 750 m^2 of hybrid maize capable producing 4.5 ton/ha (application load of 80 ton/ha). However, if P-availability is limiting, the cultivatable area is approximately 300 m^2 [38]. Apart from the nutrients recovered through the composting process, vermicomposting also allows for the cultivation of worms that can be sold to start-up other vermicomposting sites or as protein sources for animal feeds. In one year time the amount of worms can increase by a ten-fold, and one kilogram of

Table 8. Overview of technical systems for concept scenario 3—Concept of natural resources flows.

Concept scenario 3—Centralisation			
Water supply	Wastewater	Agriculture	Solid waste
- Connection to centralized water supply	- Connection to centralized sewer system and treatment	- Use of compost in local gardens/agriculture	- Separation of organic waste - Centralized composting of organic waste

worms (Eisenia fetida) will sell for about € 15 to €60, depending on the demand and quality [39].

The different technologies as well as different resources recovered are summarized in Table 9.

10.5 DISCUSSION

The participatory planning approach combining scenario development (using a Future Workshop) with participatory planning has shown to be a useful approach to raise awareness and interest of users and stakeholders to the issues at stake. Complex technical issues could successfully be discussed with a wide range of local users and stakeholders.

When approaching the water supply, sanitation, wastewater treatment and solid waste management from the resource conservation and reuse point of view a different approach to the water chain and a different interpretation of the characteristics of the water and waste is required. By following this reverse water chain approach the potential of pollutants can be seen as valuable resources, both in terms of physical resources (nutrients or water), but also as economic resources. The brief descriptions of three different decentralized technologies show that the implementation of all

Table 9. Summary of examples of on-site technologies and potential resources recovered.

Technology	Size	Potential resource recovery per household per year	Benefit
On-site rainwater harvesting with post-treatment	Roof: 36 m^2 Storage tank: 1.5–2.0 m^3	Water: 20.3 m^3 (70% efficiency)	49% reduction in water demand
EcoSan and biofiltro	EcoSan: 1.5 m^2 Biofiltro: 2 m^2	N: 5.7 kg P: 1.5 kg K: 2.4 kg Flush water: 51 m^3 Treated grey water: 31 m^3 (75% efficiency)	300–750 m^2 of maize, producing an estimated 4.5 tons/ha Redundant flush water: 51 m^3 Recovered grey water accounts for 60% of the water demand
(Vermi-)composting	Composting structure: 1m^3	N: 6 kg P: 1 kg OM: 200 kg Worms: 10 kg from 1 kg/year	300–750 m^2 of maize, producing an estimated 4.5 ton/ha

three technologies can result in an improvement of the living conditions of the inhabitants of peri-urban areas while there are also economic savings as less water or fertilizers will have to be bought and one will need to pay less to a centralized body for the treatment of one's water or wastes. In addition, the individual technologies presented require little space (1–3 m2) and hence generally fit within the gardens of the local houses, even if all three are constructed at one house although this can differ per case, and can be constructed with materials available locally in México. The fact that there is no need for large excavation for pipelines or external energy sources for pumps and regulators is another technological advantage of the decentralized technologies because it increases its versatility by making it applicable for areas not easily accessible and lessens construction costs of large sewage systems.

The implementation of decentralized technologies does not by definition mean that this is on a household, or on-site, scale. If the local context allows for this, technologies could also be implemented on communal level and thereby create small economies of scale and density with all benefits (and constraints) of this. Another advantage of decentralized technologies is that different technologies can be combined on different scales to achieve different objectives. The example presented earlier considered three specific technologies, but the research also showed that, for instance, the implementation of constructed wetlands on communal level can provide water suitable for supplementary irrigation while the digestion of organic wastes at household level provides local energy for cooking, lighting or heating. The sludge can be used as a natural fertilizer and soil conditioner, and hence an impulse to the local economy through agricultural stimulation is created.

However, as mentioned before, the implementation of the decentralized technologies and reuse of resources will require the acceptance of the technologies by society as well as a change in perception and behavior from society. In addition, it might initially cost them money as they will need to construct the different technologies. It is not self-evident that the use of collected rainwater and treatment prior to consumption (chlorine or UV-light if there is an energy source), as well as the use of treated wastewater for non-potable purposes is accepted and implemented readily by all inhabitants. Furthermore, the use of an EcoSan toilet could be perceived as less comfortable, easy or less hygienic than a conventional flush toilet. As a decentralized technology will require operation and maintenance by the inhabitant, one is directly responsible for the wastes that one produces. Hence decentralized technologies demand an active involvement of the users, for instance in the form of cleaning the roof and rainwater storage tanks, handling the (dried) feces and urine, separating wastes and periodically over-turning the compost pile. This active involvement often has a low social acceptance if there is an easier option, such as a centralized, conventional, system, available. Thus, the society plays an important role in implementing decentralized technologies as their cultural perceptions, values and wishes will need to be taken into account while designing the technology or combination of technologies, which is not self-evident [40,41].

Another obstacle for the implementation of decentralized technologies and the reuse of resources could be institutional constraints or legislation. Legislation could restrict the use of human excreta as a natural fertilizer or pose restrictions on the quality of the effluent if it is to be discharged to a natural water body. In addition, the implementation of centralized technologies can be used by authorities to gain control over resources flows to an area and hence the ability to influence its population. In the case of decentralized technologies this is much less of a possibility, and hence authorities could be less inclined to promote the approach proposed here.

The last obstacle discussed in this paper is the threat of the technologies to the public health. Although the technologies are envisaged to be implemented in peri-urban areas where there is inadequate access to safe water supply, wastewater treatment and solid waste management services, and thus improve the living conditions of the inhabitants, improper construction, operation or maintenance can result in serious health risks. This is especially the case for EcoSan, which if not constructed, operated or maintained properly can become hygienically unsafe and result in a source of pathogens or vectors, thereby increasing the risk of infections.

10.6 CONCLUSIONS

As the research results have shown, the implementation and combination of decentralized technologies allows for context-specific solutions to local water supply, sanitation or solid waste dilemmas that are often relevant in peri-urban areas. Moreover, it provides the opportunity to lessen environmental pollution and recover valuable resources to sustainably improve livelihoods and the environment. However, in order to achieve this, participation of users and stakeholders is crucial. As the experiences of this project have shown, stakeholders still often favor traditional centralized solutions and, in order to facilitate the implementation of decentralized solutions, an open and transparent planning process that highlights advantages and possible disadvantages of decentralized over centralized solutions needs to be pursued. At the very beginning of such a planning process a clear vision of the local stakeholders is necessary establishing what the existing problems are and how they could be solved. A participatory planning process linked to the identification of development scenarios (using Future Workshops) as tested in this study can help to develop technical concept scenarios that show all options for problem solving. As a next step, these options need to be carefully assessed not only from an environmental, but also from an economic and social, perspective.

REFERENCES

1. Coles, J.W.; Hesterly, W.S. Transaction costs, quality, and economies of scale: Examining contracting choices in the hospital industry. J. Corp. Financ. 1998, 4, 321–345.
2. Graham, D.J.; Couto, A.; Adeney, W.E.; Glaister, S. Economies of scale and density in urban rail transport: Effects on productivity. Transp. Res. Part E Logist. Transp. Rev. 2003, 39, 443–458.
3. Kraus, M. Economies of scale in networks. J. Urban Econ. 2008, 64, 171–177.
4. Peter-Varbanets, M.; Zurbrügg, C.; Swartz, C.; Pronk, W. Decentralized systems for potable water and the potential of membrane technology. Water Res. 2009, 43, 245–265.
5. Körner, I.; Saborit-Sánchez, I.; Aguilera-Corrales, Y. Proposal for the integration of decentralized composting of the organic fraction of municipal solid waste into the waste management system of Cuba. Waste Manag. 2008, 28, 64–72.
6. Mankad, A.; Tapsuwan, S. Review of socio-economic drivers of community acceptance and adoption of decentralised water systems. J. Environ. Manag. 2011, 92, 380–391.
7. Massoud, M.A.; Tarhini, A.; Nasr, J.A. Decentralized approaches to wastewater treatment and management: Applicability in developing countries. J. Environ. Manag. 2009, 90, 652–659.
8. Zurbrügg, C.; Drescher, S.; Rytz, I.; Maqsood Sinha, A.H.; Enayetullah, I. Decentralised composting in Bangladesh, a win-win situation for all stakeholders. Resour. Conserv. Recycl. 2005, 43, 281–292.
9. Huibers, F.P.; van Lier, J.B. Use of wastewater in agriculture: The water chain approach. Irrig. Drain. 2005, 54, 3–9.
10. Van Lier, J.B.; Huibers, F.P. From unplanned to planned agricultural use: Making an asset out of wastewater. Irrig. Drain. Syst. 2010, 24, 143–152.
11. Nanninga, T.A. Helophyte Filters, Sense of Non-sense? A Study on Experiences with Helophyte Filters Treating Grey Wastewater in The Netherlands. Master's Thesis, Wageningen University, Wageningen, The Netherlands, July 2011.
12. Merlín-Uribe, Y.; Contreras-Hernández, A.; Astier-Calderón, M.; Jensen, O.P.; Zaragoza, R.; Zambrano, L. Urban expansion into a protected natural area in

Mexico City: Alternative management scenarios. J. Environ. Plan. Manag. 2012. in press.
13. Wigle, J. The "Xochimilco" model for managing irregular settlements in conservation land in Mexico City. Cities 2010, 27, 337–347.
14. Genovevo, J.P.E. Chinampas: Entre apantles y acalotes (in Spanish). In Los Pueblos Originarios de la Ciudad de México: Atlas Etnografico; Mora, T., Ed.; Gobierno del distrito Federal and Intitutio Nacional de Antropología e Historia: Mexico City, México, 2008; pp. 97–101.
15. Martínez-Ruiz, J.L. Manual de Construcción de Chinampas (in Spanish); Instituto Mexicano de Tecnología del Agua (IMTA): Cuernavaca, México, 2004.
16. Jimenéz-Osorina, J.I. Past, present and future of the chinampas. In Maya Sustainability Report; Dry Lands Research Institute, University of California: Riverside, CA, USA, 1990.
17. González-Pozo, A. Las Chinampas de Xochimilco al despuntar el siglo XXI: Inicio de su Catalogación (in Spanish); Universidad Autónoma Metropolitana: México City, México, 2010.
18. The Ramsar Convention on Wetlands Homepage. The Ramsar list of Wetlands of International Importance. Available online: http://www.ramsar.org/cda/en/ramsar-documents-list/main/ramsar/1-31-218_4000_0__ (accessed on 27 June 2012).
19. Martínez-Ruiz, J.L. Origen, grandeza y decadencia de la chinampa: Crónica de un colapso socioambiental anunciado (in Spanish). In Proceedings of the 12th National Irrigation Congress (Congreso Nacional de Irrigación), Zacatecas, México, 13–15 August 2003; ANEI 2003 M6 T14. pp. 107–116.
20. Programa de Desarrollo Urbano de la Delegación Xochimilco (in Spanish); Secretaria de Desarrollo Urbano y Vivienda: Mexico City, México, 2005.
21. Bernstein, J. Toolkit for Social Assessment and Public Participation in Municipal Solid Waste Management; Urban Environmental Thematic Group, World Bank: Washington, DC, USA, 2004.
22. Bromell, D.; Hyland, M. Social Inclusion and Participation: A Guide for Policy and Planning; Ministry of Social Development: Wellington, New Zealand, 2007.
23. FAO. Participatory Communication: A Key to Rural Learning Systems; Food and Agricultural Organization: Rome, Italy, 2003.
24. Jungk, R.; Müller, N. Future Workshops: How to Create Desirable Futures; Institute of Social Inventions: London, UK, 1987.
25. Qué Sabemos de Orina Humana Como Fertilizante—What we know about human urine as a fertiliser (in Spanish). Available online: http://www.sarar-t.org (accessed on 11 June 2012).
26. Gener Cada Mexicano Un Kilogram De Baura El Día. México: Noticias 8 May 2011 (In Spanish). Available online: http://noticias.universia.net.mx/en-portada/noticia/2011/08/05/854479/genera-cada-mexicano-kilogramo-basura-dia.html (accessed on 18 June 2012).
27. Enviros RIS. Residential Waste Materials Density Study; No. WDO OPT/ORG-R2-02; Waste Diversion Organization: Toronto, Canada, 2001. Available online: http://www.wdo.ca/files/domain4116/Access%20OPT_ORG%20R2-02.pdf (accessed on 18 September 2012).

28. Jönsson, H.; Baky, A.; Jeppsson, U.; Hellström, D.; Kärrman, E. Composition of Urine, Faeces, Greywater and Biowaste for Utilisation in the Urware Model; Urban Water, Chalmers University of Technology: Göteborg, Sweden, 2005.
29. Normales Climatológicos 1971–2000, Estación San Gregorio Atlapulco—Climatological normals 1971–2000, San Gregorio Atlapulco Station (in Spanish) Available online: http://smn.cna.gob.mx (accessed on 27 June 2012).
30. Instituto Mexicano de Tecnología del Agua (IMTA). Personal communication. Personal Communication with Geographer during a workshop in San Gregorio Atlapulco (Xochimilco) for the Project Vivace, Instituto Mexicano de Tecnología del Agua (IMTA): Cuernavaca, México, 2010.
31. Instituto Mexicano de Tecnología del Agua (IMTA). Personal communication. Personal Communication with Expert and Survey Conductor, Instituto Mexicano de Tecnología del Agua (IMTA): Cuernavaca, México, 2012.
32. Instituto Mexicano de Tecnología del Agua (IMTA). Detailed Household Surveys Conducted in La Conchita and San Martín (Xochimilco) for the Project Vivace; Instituto Mexicano de Tecnología del Agua (IMTA): Cuernavaca, México. 2011.
33. Belmont, M.A.; Cantellano, E.; Thompson, S.; Williamson, M.; Sánchez, A.; Metcalfe, C.D. Treatment of domestic wastewater in a pilot-scale natural treatment system in central México. Ecol. Eng. 2004, 23, 299–311.
34. Breitenbeck, G.A.; Schellinger, D. Calculating the reduction and material mass and volume during composting. Compost Sci. Util. 2004, 12, 365–371.
35. Yue, B.; Chen, T.; Gao, D.; Zheng, G.; Liu, B.; Lee, D. Pile settlement and volume reduction measurement during forced-aeration static composting. Bioresour. Technol. 2008, 99, 7450–7457.
36. Tilley, E.; Lüthi, C.; Morel, A.; Zurbrügg, C.; Schertenleib, R. Compendium of Sanitation Systems and Technologies; Swiss Federal Institute of Aquatic Science and Technology (Eawag): Dübendorf, Switzerland, 2008.
37. Dierolf, T.; Fairhurst, T.; Mutert, E. Soil Fertility Kit: A Toolkit for Acid, Upland Soil Fertility Management in Southeast Asia; Singapore Potash & Phosphate Institute (PPI): Singapore, 2001.
38. Richert, A.; Gensch, R.; Jönson, H.; Stenström, Dagerskog, L. Practical Guidance on the Use of Urine in Crop Production; EcoSanRes Series 2010-1; Stockholm Environmental Institute (SEI): Stockholm, Sweden, 2010.
39. Gómez, L.M. Manual De Lombricomposta (in Spanish); Secretaría de Educación Publica (SEP) and Colegio de Estudios Cientificos y Tecnológicos del Estado de Chiapas: Tuxtla Gutiérrez, México, 2008.
40. Martínez-Ruiz, J.L.; Murillo, D. The problem of the social adoption of appropriate technologies. In Revista Plurimondi: An International Forum for Research and Debate on Human Settlements; Edizioni Dedalo: Bari, Italy, 2010; Volume 4, pp. 125–145.
41. Starkl, M.; Bisschops, I.; Norström, A.; Purnomo, A.; Rumiati, A. Integrated assessment of the feasibility of community based sanitation options: A case study from East Java, Indonesia. Water Pract. Technol. 2010, 5, 1–12.

CHAPTER 11

Global Analysis of Urban Surface Water Supply Vulnerability

JULIE C. PADOWSKI AND STEVEN M. GORELICK

11.1 INTRODUCTION

Rapid population growth over the past 60 years has led to accelerated global urbanization. With more than half of the world's inhabitants now residing in cities, the human geography of the planet has shifted, condensing nearly 3.6 billion people into <1% of the total global land area (Schneider et al 2009, United Nations 2012). These urbanized areas represent hotspots of localized, concentrated development that place enormous pressure on natural resources, particularly freshwater supply. Globally, urban water use has increased five-fold since 1950, reflecting greater demands as more people move to urban centers and increase their per capita use, commensurate with higher standards of living (Richter et al 2013). Combined with water demands for agricultural irrigation, industrial production, and water to sustain environmental flows, water scarcity has become a global problem.

The impact of urban demands on freshwater resources not only alters the hydrologic balances in the basins that supply and receive water for urban areas, but can devastate the integrity of a watershed tapped for supply (Miltner et al 2004). An historical example is the city of Los Angeles, US, which in the early 1900's commandeered nearly the entire flow of the Owens Valley and

Mono Lake basins, severely damaging the existing natural and human systems within each. More modern examples include the dramatic decline in the water quantity and quality of Lake Chapala through the early 2000's caused in part by the city of Guadalajara, Mexico drawing increasingly larger volumes of water for urban supply (Webster et al 2009). The Krishna basin in India suffers "significant degradation of various ecosystems" due to widespread over-allocation across all sectors, including large withdrawals for the cities of Hyderbad, Pune, and water exports to Chennai (Venot 2009). On the other hand, places like New York City, US have instead promoted source watershed protection measures that leave supply basins with functional ecosystems, despite providing water to one of the largest cities in the world.

Freshwater vulnerability is only partially a reflection of growing urban demands. Although water for municipal use has become significantly more important over the past century, agricultural irrigation is by far the dominant user of water around the world (Gleick 1993). Since the 1950's, the volume of water used for irrigation purposes has nearly tripled (Molden 2007). However, historical trends suggest that as urbanization intensifies, a larger portion of available freshwater resources will be channeled to the urban sector (Oki and Kanae 2006). The options for closing the growing urban water deficit are building non-traditional, expensive sources (e.g., recycling, desalination), developing reallocation schemes, such as urban-rural partnerships (Gober 2010, Richter et al 2013), or by reducing water allocated for environmental flows (Rogers 1993).

Until recently, urban impacts on supply vulnerability were either performed for individual cities, regionally, e.g., in the US (Averyt et al 2013, Padowski and Jawitz 2012), or by assuming the basins within which urban areas reside are those supporting urban water demands (Jenerette and Larsen 2006, McDonald et al 2011). Currently, only the City Water Map Initiative has presented a global assessment of urban water stress that directly incorporates information about urban water sources (McDonald et al 2014).

The study presented here examines baseline and future vulnerability scenarios of cities under normal climate conditions and quantifies the magnitude of reallocation needed to reduce future vulnerability. To accomplish this, our global analysis evaluates the vulnerability of surface water supply basins serving large cities that lack source diversity, relying on either direct river withdrawals or releases from reservoir storage and asks: 1) Which large urban areas are most vulnerable to demands for freshwater now and in the near future? 2) Of those vulnerable large cities, what is the magnitude of water reallocation needed to assure urban freshwater supplies?

11.2 METHODOLOGY

11.2.1 CRITERIA FOR INCLUSION AND DATA SOURCES

Water vulnerability assessments were performed for all basins supporting cities with populations in excess of 750 000 that obtain water supplies solely from either (a) river withdrawals or (b) reservoirs (natural or man-made) that are not shared with another large city. In total, 70 cities in 39 countries were identified that meet these criteria. Information on the type and location of urban water sources was obtained from the City Water Map Initiative (McDonald et al 2014). Hydrologic data, including mean annual reliable basin discharge (GWSP 2008b), environmental flow requirements (GWSP 2008a), mean annual reservoir storage capacity (Lehner et al 2011, NID 2009), global potential evapotranspiration rates (Zomer et al 2007, 2008) and urban demand (FAO 2013) were obtained from global databases (see supplemental material). Available environmental flow requirements and discharge statistics for all supply basins were based on time series of monthly climate variables over the climate normal period 1961-1990. Estimates of future vulnerability are based on urban growth projections developed in the UN World Urbanization Prospects (United Nations 2012). Basin boundaries for each urban water source were delineated using HydroSHEDS (Lehner et al 2008), yielding 106 surface-water basins, with 48 cities relying on a single supply basin and 22 cities relying on more than one basin. Where a city uses more than one supply basin, that city's supply basin refers to the combination of discharge from all of the supply basins used.

11.2.2 OBJECTIVES

The focus of this study is on large cities using surface water supplies that lack source diversity, relying on either direct river withdrawals or releases from reservoir storage. Three vulnerability metrics, discussed below, were used to identify those cities that fail to meet simultaneous thresholds of freshwater supplies for urban, agricultural, and environmental purposes. Urban water vulnerability was calculated for baseline conditions (2010) and for a future scenario (2040) in which increasing water demands are proportional to predicted population growth and projected irrigated agricultural demand. Although projected demand estimates account for population increases, surface water supply changes resulting from expanding urban boundaries and additional future supplies a city might tap by 2040 are not considered. These estimates also do not include any impacts of changes in land use or climate, which could potentially increase or decrease the amount of water available to an urban area (Velpuri and Senay 2013). Rather, this analysis establishes a baseline estimate of vulnerability assuming stationarity in urban hydrologic systems over the next 25 years. For the future scenario, the reduction in urban water vulnerability that could be achieved is examined in two end-member demand scenarios: 1) demands based on projected water-sharing allocations, and 2) demands where excess water from agriculture or the environment is redirected for the other sectoral uses.

11.2.3 BASIN VULNERABILITY METRICS

11.2.3.1 ENVIRONMENTAL WATER REQUIREMENTS (Q_E)

Environmental flows are a critical water resources component (Poff et al 1997, Smakhtin et al 2004). Ideally, a supply basin should be capable of fostering urban well-being while supporting a viable ecosystem. The environmental water metric reflects the vulnerability of freshwater-dependent ecosystems to urban water use, and is defined as a ratio of critical environmental water demand to available basin flow. Environmental water demand (D_E) is based on low-flow conditions derived from the WaterGap model in which global estimates of environmental water flow were determined at the watershed scale (Döll et al 2003, Smakhtin et al 2004).

This vulnerability metric compares the water required to meet environmental demands to the mean low-flow monthly water discharge within each basin (q_{90}). Although it has been shown that urbanization can cause changes in runoff in some US urban watersheds (Velpuri and Senay 2013), global data of changes in runoff trends are currently not available and may not be applicable in cities that collect water from basins external to their urban footprint. Rather, within each basin, the q_{90} value is used to represent the "reliable monthly discharge," which is equal to the minimum flow that is exceeded on average 90% of the time (based on monthly discharge data). We define the environmental water requirement (Q_E) for each basin as the ratio of D_E to the low-flow runoff (q_{90}) after agricultural (D_A) and urban demands (D_H) are met (see 2.3.2 for agricultural and urban demand calculations) (equation (1)).

$$Q_E = D_E / (q_{90} - D_H - D_A) \qquad (1)$$

Any basins for which environmental water demands (D_E) are large relative to net reliable flow are considered threatened ($Q_E > 0.4$). The threshold of 0.4 adopted here builds on a similar scarcity threshold applied in other well-established contexts (Falkenmark et al 1989, Vörösmarty et al 2005).

11.2.3.2 HUMAN USES: URBAN WATER AND AGRICULTURAL IRRIGATION

Urban water requirements were assessed using an analogous metric to that used for environmental flow requirements. Specifically, the urban water requirement represents a ratio of urban demand during low-flow conditions to renewable surface water in basins when environmental needs are met first. Since urban water demand (D_H) is not a commonly reported metric, estimates of demand for each city were computed by multiplying the population by the mean annual national municipal and industrial per capita demands, excluding withdrawals for thermoelectric cooling (GWSP 2008c). The discharge required to meet urban needs was calculated as the ratio of D_H to the mean "reliable monthly discharge" for a basin (q_{90}) after agricultural (D_A) and environmental demands (D_E) are met (equation (2)).

$$Q_H = D_H / (q_{90} - D_E - D_A) \qquad (2)$$

As with Q_E, a threshold is applied such that all basins for which urban demands DH are large relative to net available low flow are deemed threatened ($Q_H > 0.4$).

Agricultural demands (D_A) were estimated from the most recent annual national water use statistics (FAO 2013) for urban supply basins with >1% of the total area occupied by land under irrigation (Siebert et al 2013). The relative percent of water withdrawn for agriculture was used as a proxy for the flow in each basin that is allocated for irrigation purposes. In Eq. 2, systems supported by river withdrawals, have D_A subtracted directly from the q_{90} while basins supporting urban supplies through reservoirs, versus directly from streamflows, have a D_A=0. For these reservoir-supplied systems, values of D_A are instead subtracted from reservoir storage (equation (3)). Projections of future agricultural demands for the 2040 scenario were made based on available national water use statistics from 1965-2010.

11.2.3.3 STORAGE REQUIREMENTS (Q_S)

The third measure of vulnerability provides insight into a basin's hydraulic capacity to supply water. This measure is based on the ratio of total annual urban demand volume to reservoir normal storage. Reservoir storage is a key component in many urban supply systems, increasing hydraulic system stability by helping to buffer variability in available supply (Padowski and Jawitz 2012). In this study, 28 of the 70 cities relied on releases from surface reservoirs. Here, urban supply basin reservoirs are assumed to be managed such that urban demands receive top priority. It is also assumed that no additional significant reservoir capacity will be added before 2040. Understanding that these assumptions likely provide just one assessment of water available for human consumption, storage threat is measured as the ratio of urban water demand (D_H) to the reported reservoir annual storage capacity (Q_{cap}) minus evaporative losses estimated using potential evapotranspiration (D_V) (Ward and Trimble 2003) and agricultural demands (D_A) for a year (equation (3)), where basins are considered threatened when Q_S>0.4 (Lane et al 1999).

$$Q_S = D_H / (Q_{cap} - (D_V + D_A)) \qquad (3)$$

11.2.4 CITIES SUPPLIED BY DIRECT RIVER WITHDRAWALS VERSUS RESERVOIRS

Reservoir-supplied cities were evaluated using each of the three metrics described above, however river-supplied cities have no major storage infrastructure and were assessed using only the two flow-based metrics (Q_E and Q_H).

11.2.5 VULNERABLE VERSUS THREATENED STATUS

For both categories—reservoir and river supplied—cities were considered "vulnerable" if all demand thresholds were violated. If a city had one or two but not all three metrics exceeding a threshold, then it was considered "threatened," as these supplies may no longer be able to support urban demand without causing damage to agriculture, the environment, or both. We refer to a city as "susceptible" if it has either threatened or vulnerable status; the number of susceptible cities is the sum of those that are threatened or vulnerable. Finally, cities were considered "non-threatened" if no demand thresholds were violated.

11.3 RESULTS

11.3.1 BASELINE VULNERABILITY AND THREATS (2010)

Each city was assessed according to the three vulnerability metrics defined in section 2. The status of each city, either vulnerable (dark), threatened (white) or non-threatened (light gray) is shown on the global map in figure 1. The total number of cities in each category is given in table 1. Of the 70 cities included in this study, 25 are classified as vulnerable in 2010, of which 23 are cities supplied solely by direct river withdrawals. These vulnerable river-supplied cities have significantly lower mean per capita flows (2700 l/p/d) than those cities with sufficient water (200 200 l/p/d), and fall within the range of water availability that Falkenmark et. al

(1989) identify as having water management problems due to water stress (2700–4700 l/p/d).

In addition to supporting urban demands, urban supply basins often provide water for agricultural and environmental needs as well. To better understand why vulnerability occurs, the primary source of water demand by sector was determined for each supply basin. Considering only the 25 vulnerable cities, demand from the agricultural sector is dominating for the supply basins of 16 cities (64%). There are five vulnerable cities for which environmental needs are most prominent within supply basins (20%), with demand from the urban sector dominating in the supply basins of four cities (16%) (table 2).

Also examined were 'threatened' cities, where 'threatened' means that one or two of the three metrics exceed their demand threshold. Of the 70 cities considered, 33 are threatened and represent the majority of reservoir-supplied systems (71%), but only 31% of river-supplied cities are threatened (table 1). Environmental demand was the only stressor for river-supplied cities, however reservoir-supplied cities revealed more varied sources of stress, suffering from a variety of environmental, human and storage-related threats (table 2). The number of threatened reservoir-supplied cities is ten times greater than those considered vulnerable, whereas the number of threatened river-supplied cities is approximately half of those found to be vulnerable (table 1).

11.3.2 FUTURE VULNERABILITY AND THREATS (2040)

In 2040, the number of vulnerable cities rises from 25 to 31, representing a 24% increase in urban areas facing serious water issues (table 1). Of these 31 cities, those that are river-supplied continue to make up the majority of those that are vulnerable. Our analysis provides a baseline estimate of vulnerability assuming near-term stationarity in urban surface water hydrologic systems. Even without accounting for potential changes in local hydrology due to climate change, as populations increase, mean per capita flows for vulnerable river-supplied cities drop over this 30 year period to 1200 l/p/d, creating "chronic water scarcity" (Falkenmark et al., 1989). During this same period, the total number of vulnerable reservoir-supplied

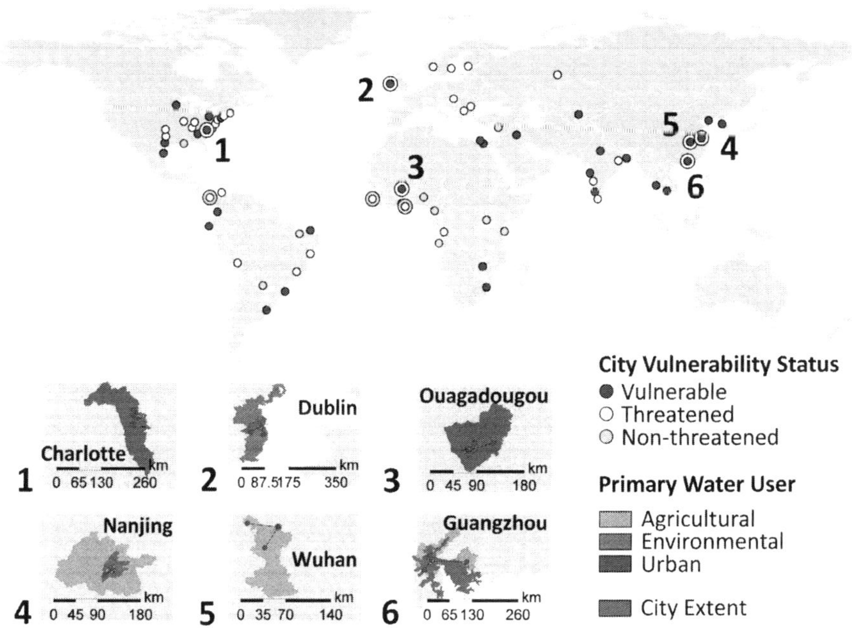

Figure 1. City supply system vulnerability. Cities are represented on the world map by circles where shade indicates the degree to which each city faces water supply issues. Cities that change status over the time period studied are demarcated with a dark outer ring (2010) and inner circle (2040). Vulnerable cities (dark) are those who suffer from both environmental and urban water scarcity. Threatened cities (white) are those that have either environmental or urban water scarcity issues. Cities with no water supply issues (non-threatened) are symbolized with light gray. Numbered locations represent those cities-of-concern (see section 3.3) which become vulnerable by 2040. Insets of these particular cities accompany the world map. Dark circles represent the general location of urban supply sources in inset maps. Basin shading and pattern denotes the primary sectoral water user in each system, whereas dark gray areas indicate city extents. City names may be found in the supplementary information.

cities more than doubles, from 7% to 18%, as storage infrastructure loses its advantage as a supply buffer. When considering only the 31 vulnerable cities in 2040 (table 2), 17 (55%) cities had supply basins dominated by demand from the agricultural sector. The number cities with vulnerable supply basins dominated by demand from the urban sector increases from four (16%) in 2010 to eight (26%) in 2040.

Although more cities become vulnerable in 2040, the percent of 'threatened' cities decreases from 47% in 2010 to 43% in 2040 (table 1). Similar to 2010, the majority of reservoir-supplied cities are those considered threatened (71%), however the relative difference between the numbers considered threatened versus vulnerable decreases. Environmental demands continue to be the only stressor for river-supplied cities, however, stress from storage over-allocation becomes the leading threat in reservoir-supplied cities, occurring in 15 of the 28 cities depending on this supply type (table 2). Although the ratio of reservoir- to river-supplied threatened cities increases, by 2040 all but three reservoir-supplied and six river-supplied cities remain unthreatened. Access to large sources of water combined with relatively low urban demand and/or environmental demands keeps these nine cities protected from water-supply compromise.

11.3.3 "CITIES-OF-CONCERN"- LOCATIONS THREATENED OR VULNERABLE IN THE NEAR FUTURE

Between 2010 and 2040, nine "cities-of-concern" were identified as becoming more susceptible, with conditions worsening from a non-threatened to threatened status, or from threatened to vulnerable. In total, three of the cities identified as non-threatened in 2010 become threatened in 2040 (Accra, Ghana; Freetown, Sierra Leone; Panama City, Panama). Six cities-of-concern change from threatened to vulnerable status (Dublin, Ireland; Charlotte, USA; Ouagadougou, Burkina Faso; Guangzhou, Wuhan, and Nanjing, China; figure 1— insets). Three of these cities-of-concern (Charlotte, Ouagadougou, and Panama City), each transition from a system dominated by the environmental sector demand to one dominated by the urban sector demand, as none of their supply basins support significant irrigated agriculture. Of these nine total cities that show increased susceptibility by 2040, six are reservoir-supplied cities and three are

TABLE 1. Number of large cities either vulnerable or threatened.

Year	Type	Number of Cities by Category			Percent of Cities by Category		
		All Urban Systems	Reservoir-supplied	River-supplied	All Urban Systems	Reservoir-supplied	River-supplied
2010	Vulnerable	25	2	23	36%	7%	55%
	Threatened	33	20	13	47%	71%	31%
	Not Threatened	12	6	6	17%	21%	14%
	Total	70	28	42	100%	100%	100%
2040	Vulnerable	31	5	26	44%	18%	62%
	Threatened	30	20	10	43%	71%	24%
	Not Threatened	9	3	6	13%	11%	14%
	Total	70	28	42	100%	100%	100%

river-supplied. This suggests that increased demand by 2040 puts at risk a significant number of both river-supplied and reservoir-supplied cities.

11.3.4 WATER REALLOCATION TO REDUCE THREATS TO CITIES IN 2040

Two water reallocation strategies are examined to assess their potential usefulness of reducing the threat of failing to meet urban demands. In these alternative assessments, urban water threat predictions operate under the assumption that growing urban demands may either completely consume water required for environmental flows (reallocation through environmental neglect), or be tempered through an urban-rural partnership where water is reallocated from agriculture via a water market to urban uses within a supply basin (reallocation through agricultural transfers). In this scenario, surface reservoirs continue to provide irrigation water and a portion of that water is transferred to the urban sector sufficient to meet the urban water demand threshold. The remaining water is used for irrigation or to satisfy environmental needs. Similarly, irrigation water in excess of urban needs is also used to meet minimum environmental flows in river-supplied cities. This analysis represents an end-member strategy in which enough irrigation water is transferred to satisfy urban demands plus other sectoral demands to the degree possible.

Table 3 shows the number of cities for which reallocation provides sufficient water under three scenarios: (1) satisfying urban demands only, (2) satisfying both urban and environmental demands, and (3) securing adequate reservoir storage. Here, "sufficient water" means that enough water exists within a city's supply basin to satisfy sectoral demands in each scenario.

Growing demand for freshwater will likely force human needs to be placed ahead of environmental requirements. As such, the first strategy is reallocation through environmental neglect in which cities divert all water available for maintaining minimum environmental flows to meet urban needs. Under this strategy agricultural demands are also met, such that neither the urban or agricultural sectors are threatened. Of the 70 cities included in this study, the number of cities that could support urban

TABLE 2. Number and percent of vulnerable (upper) and threatened (lower) cities classified by dominant cause and supply type. Also identified are the number of threatened cities that face either single or dual threats to a city's water supply status.

	2010		**2040**	
Dominant Cause	Reservoir-supplied	River-supplied	Reservoir-supplied	River-supplied
Vulnerability dominated by demand from:				
Agriculture	2	14	1	16
Environment	0	5	1	5
Urban	0	4	3	5
Total Vulnerable (% by supply type)	2 (7%)	23 (55%)	5 (18%)	26 (62%)
Threat dominated by stress from:				
Environment	5	13	3.5	10
Human	1	0	1.5	0
Storage	14	---	15	---
Total Threatened (% by supply type)	20 (71%)	13 (31%)	20* (71%)	10 (24%)
Sole Threat (% by supply type)	11 (55%)	13 (100%)	10 (50%)	10 (100%)
Multiple Threats (% by supply type)	9 (45%)	0 (0%)	10 (50%)	0 (0%)

* The total number of threatened reservoir-supplied cities in 2040 includes fraction values because the dominant threat in one city is tied - both environmental and human threat scores are equal.

demands in 2040 increases from 35 to 51 when exercising environmental neglect reallocation (table 3). Urban sector threats alone are reduced in 13 otherwise vulnerable and 29 threatened cities, where nine cities remain non-threatened. Those aided by reallocation through environmental neglect include four of the six cities-of-concern likely to become vulnerable by 2040, with only Ouagadougou and Charlotte failing to benefit.

The second strategy reallocates water from agriculture to urban use and shows similar benefits to reallocation by environmental neglect for 50 versus 51 cities. However, agricultural reallocation preserves environmental flows, and shows that roughly the same benefit can be gained without further damage to freshwater-dependent ecosystems. Under this strategy, urban sector threats are reduced in 15 otherwise vulnerable and 26 threatened cities, whereas nine cities remain non-threatened. Reallocation from agriculture leaves 29% of cities (including Ouagadougou, Charlotte, and Dublin) unable to meet urban demands. Additionally, under reallocation of irrigation water, the number of cities for which both urban and environmental demands are met increases from 18 to 28, and the number able to achieve reservoir storage security increases from 10 to 13 cities (table 3). Water available only from agricultural transfers could only support both urban and environmental demands in 3 of the 10 cities-of-concern (Guangzhou, Nanjing, Wuhan) that become more vulnerable between 2010 and 2040. A combination of environmental neglect and agricultural transfers would leave only Ouagadougou vulnerable in 2040.

11.4 CONCLUSIONS

Cities here are considered 'vulnerable' when they exceed minimum thresholds for human, environmental, and storage requirements. Those cities exceeding some but not all three thresholds are considered 'threatened'. The nature of urban vulnerability discussed here is exclusively focused on large cities judged most at risk for water scarcity because they lack source water diversity. By 2040, 31 of 70 large cities relying on surface water are predicted to become vulnerable. This represents an increase from 36% vulnerable in 2010 to 44% in 2040. In the majority of cases, the number of cities unable to meet human-based demands is likely to outstrip the number of cities for which minimum flow requirements of freshwater-dependent ecosystems are not met. The number of vulnerable river-supplied cities greatly exceeds reservoir-supplied cities in 2010 (tenfold) and 2040 (fivefold). Of large surface-water dependent cities in 2010, 83% are either vulnerable or threatened. This increases to 87% in 2040.

Although the majority of threatened cities are reservoir-supplied, artificial storage currently offers far more security to cities using reservoirs. This is indicated by the substantially lower number of vulnerable reservoir-supplied (18%) versus river-supplied cities (62%) in 2040. In particular, reservoirs may have allowed some cities to delay vulnerability in the near future, but the high number of threatened cities suggest that the large financial investments made in the past to improve water security through reservoir storage may no longer be sufficient to prevent vulnerability. It is surprising that reservoir-supplied cities account for 6 of the 9 cities-of-concern showing an increase in vulnerability or threat status (susceptible cities) by 2040. This suggests that the temporal buffering capacity will no longer be adequate to meet both increasing urban and agricultural demands as well as accommodate environmental needs.

Large cities can meet urban demand via reallocation by 2040. Reallocation through environmental neglect enables 51 of 70 cities to meet urban demand, and transfers of irrigation water enable 50 cities to meet urban demand. This compares favorably to just 35 cities meeting urban demand when no reallocation exists. Agricultural transfers have the benefit of lessening urban sector threats while reducing the number of cities experiencing environmental and storage threats. However, reallocation of irrigation water by 2040 can enable both urban and environmental demands to be met in only an additional 14% of cities. Reallocation by environmental neglect enables an additional 23% of cities to meet demand. Of the nine cities-of-concern becoming vulnerable by 2040, only Ouagadougou, Burkina Faso, does not benefit from any reallocation method—even if both environmental and agricultural supplies are both transferred. Source diversification is essential in this and other regions.

These results offer much-needed insights into urban supply vulnerability, but are subject to limitations that accompany the majority of data mining studies. In using pre-existing databases from multiple sources, these analyses accommodate differences in data quality and coverage (e.g., little

TABLE 3. Number and percent of cities meeting demand and cities that are vulnerable under three reallocation strategies.

Strategy	Demand met- No Reallocation	Demand met- After Reallocation	Remaining Vulnerable
Reallocation through environmental neglect to satisfy			
Urban demand only	*35 (50%)*	*51 (73%)*	*19 (27%)*
Reallocation through agricultural transfers to satisfy			
Urban demand only (% of all cities)	35 (50%)	50 (71%)	20 (29%)
Urban & environmental demand (% of all cities)	18 (26%)	28 (40%)	42 (60%)
*Storage security (% of all cities)**	*10 (14%)*	*13 (19%)*	*15 (21%)*
Reallocation through combined environmental neglect and agricultural transfers to satisfy			
Urban demand only (% of all cities)	35 (50%)	57 (81%)	13 (19%)

* Percent totals for reallocation through agricultural transfers to satisfy storage security do not add to 100% since there are only 28 cities with reservoir storage.

data on urban water supply and demand information at appropriate spatial and temporal resolutions), which undoubtedly introduce error and restrict quantification of uncertainty. As such, these analyses only provide insights into the likely trends in urban supply vulnerability. Further investigations of how adaptations to water scarcity and climate change will impact both river- and reservoir-supplied cities would greatly add to our understanding of urban water vulnerability, and are necessary for urban water managers pursuing policies aimed at long-term water supply sustainability.

11.5 ACKNOWLEDGMENTS

This chapter is a reproduction of the ERL article corrected with material appearing in the corrigendum (Padowski and Gorelick 2014 ERL 9 104004; corrigendum 2014 ERL 9 119501. CC BY). This research was supported by the Global Freshwater Initiative through Stanford's Woods Institute for the Environment. The authors would like to thank Rob McDonald and the City Water Map Initiative team for access to their database. This work was support in part by the National Science Foundation under grant GEO/OAD-1342869 to Stanford University. Any opinions, findings, and conclusions or recommendations expressed in this material are those of the authors and do not necessarily reflect the views of the National Science Foundation.

11.6 SUPPLEMENTAL MATERIALS

SECTION S-1: DESCRIPTION OF DATA USE AND AVAILABILITY

S-1.1 URBAN SOURCE LOCATIONS

CITY WATER MAP

The City Water Map is a global survey of water sources for large cities with populations greater than 750,000 in 2010. This database identifies the names and spatial coordinates of the primary water sources and water providers for 265 urban areas. More detailed information on the City Water Map is available within the supplementary data found in McDonald et al. 2014.

HYDROSHEDS

(http://hydrosheds.cr.usgs.gov/) HydroSHEDS is a high-resolution database of hydrographic information that provides information on river networks, watershed boundaries, drainage directions and flow accumulations at various scales. For this study, the locations of urban water sources were intersected with HydroSHEDS watershed boundaries to identify the relevant urban watersheds of interest.

S-1.2 DISCHARGE AND FLOW REQUIREMENT STATISTICS

MEAN ANNUAL RELIABLE BASIN DISCHARGE

(http://atlas.gwsp.org/index.php?option=com_content&task=view&id=55&Itemid=68) Reliable mean annual water discharge was estimated from a WaterGAP dataset of reliable monthly water discharge. This dataset presents a gridded global assessment of accumulated runoff (surface runoff plus groundwater recharge). Due to often large temporal variability in river discharge over time, long-term averages may not be particularly good indicators of the volume of water that can be reliability accessed each year. WaterGAP presents the "90% reliable monthly discharge" as an alternative to long-term mean annual discharge. Estimates are "a statistical estimate of the minimum monthly flow which occurs over 90% of the months during the climate normal period (1961-1990)". These data were used to estimate the mean annual reliable discharge in the basis from which urban areas acquire their water supplies.

ENVIRONMENTAL FLOW REQUIREMENTS

(http://atlas.gwsp.org/index.php?option=com_content&task=view&id=191&Itemid=63) The data used to estimate the volume of water required to maintain freshwater-dependent ecosystems was based on the work done by Smakhtin et al. 2004. This dataset provides a spatially distributed estimate of the fraction of mean annual flow required to maintain ecosystem in "fair condition", where "fair" implies that the ecosystem endures some level of disturbance such that some sensitive species may have been lost and/or reduced in numbers in these areas. This category assumes that human disturbances are present (e.g. dams, water diversions and water quality degradation). Assumptions made to quantify the volume of water needed to maintain ecosystems in this condition are based on estimates of appropriate low- and high-flow requirements. Futher information on how these flow requirements were calculated can be found in Smakhtin et al. 2004.

S-1.3 RESERVOIR SIZE INFORMATION

GLOBAL RESERVOIR AND DAM DATABASE & NATIONAL INVENTORY OF DAMS DATABASE

(http://sedac.ciesin.columbia.edu/data/set/grand-v1-reservoirs-rev01; http://geo.usace.army.mil/nid/index.html) The GRanD database contains an extensive collection of information on nearly 7,000 dams and reservoirs around the world. Here, the GRanD database served as the primary source for information on reservoir location, storage capacity and surface area. In this study, the mean annual storage capacity of reservoirs was represented by the "normal capacity" data provided. Normal capacity in this context usually refers to the most commonly reported storage value for a given reservoir. In some cases, the normal capacity was estimated using information on dam height and reservoir surface area (see GRanD database Technical documentation for more details). Reservoir surface areas were also derived from reported GRanD values, and are assumed to be the surface area of the reservoir at normal capacity.

The National Inventory of Dams (NID) database was utilized in cases when GRanD data did not exist. The NID database contains much of the same information as GRanD, but focuses specifically on reservoirs in the

United States of America. Again, the mean annual storage capacity of reservoirs was assumed to be their "normal storage" which NID defines as "the total storage space in a reservoir below the normal retention level, including dead and inactive storage and excluding any flood control or surcharge storage". Surface area of the reservoir is given for the impoundment at its normal retention level.

In cases where neither the GRanD nor NID databases provided necessary information, the authors relied on previously published case studies, white literature, and utility websites to estimate reservoir volumes. Estimates of unknown surface areas were made from satellite images of individual reservoirs using Google Earth, and thus may not reflect the surface area at normal storage levels. Reservoir storage estimates included information from only those reservoirs believed to be used for supply (e.g. not smaller storages for local distribution of treated water).

CGIAR CSI- GLOBAL POTENTIAL EVAPO-TRANSPIRATION (GLOBAL-PET) GEOSPATIAL DATASET

(http://www.cgiar-csi.org/data/global-aridity-and-pet-database) Based on the assumption that PET is a good surrogate for lake evaporation (Ward and Trimble, 2003), this dataset was used to estimate the volume of water unavailable to urban areas using reservoirs due to evaporative demands. The Global PET dataset is a high-resolution raster that models monthly and yearly average PET using the WorldClim Global Climate database for input parameters. PET is estimated using a modified Penman-Monteith equation, developed to be particularly amenable to global analyses as it does not require estimations of site-specific parameters. In this study, the mean annual PET was measured at the dam location for each reservoir assessed. Averaged values of mean annual PET were calculated for larger water bodies, specifically the Great Lakes and Lake Victoria. Evaporative losses from reservoirs were calculated as the mean annual PET multiplied by the reported surface area of the reservoir. The volume of water lost to evaporation was then simply subtracted from the mean annual storage capacity of each reservoir.

S-1.4 WATER DEMANDS

FAO AQUASTAT

(http://www.fao.org/nr/water/aquastat/) Human Demands: We utilized the Food and Agriculture Organization's AQUASTAT database to downscale estimates of national human and agricultural water use statistics to the urban level. Data for the most recent national municipal and human withdrawals came directly from the online database for all countries included in this study. Data were converted to per capita volumes by dividing by the national population.

Agricultural Demands: Data on national agricultural water demands were also obtained from AQUASTAT. Here, the volume of water used for agriculture in urban supply basins was determined as the ratio of agricultural to total water withdrawals. This ratio was used to estimate the volume of water in each basin that is allocated for irrigation purposes in basins with >1% of land area devoted to agriculture. The total land under irrigation within each basin was determined by intersecting each urban supply watershed boundary with the gridded estimates of area equipped for irrigation provided within the AQUASTAT database.

PROJECTED ESTIMATES OF DEMAND

Estimates of human and agricultural water use in 2040 were based on available national water use statistics from 1965-2010. Projections were made using a simple exponential growth model.

WATER WITHDRAWALS FOR POWER PLANTS

(http://atlas.gwsp.org/index.php?option=com_content&task=view&id=47&Itemid=68) Estimates of mean annual water usage for thermoelectric withdrawals were made for each water supply basin included in this study. Data were obtained from a publicly available WaterGAP output found on the Global Water Supply Project website.

Table S-2.1. List of Cities (by country) and their water vulnerability status in 2010 and 2040.

City	Country	2010 Vulnerability	2040 Vulnerability
Luanda	Angola	Non-threatened	Non-threatened
Rajshahi	Bangladesh	Vulnerable	Vulnerable
La Paz	Bolivia	Threatened	Threatened
Belo Horizonte	Brazil	Threatened	Threatened
Florianópolis	Brazil	Vulnerable	Vulnerable
Fortaleza	Brazil	Vulnerable	Vulnerable
Salvador	Brazil	Threatened	Threatened
Teresina	Brazil	Non-threatened	Non-threatened
Sofia	Bulgaria	Threatened	Threatened
Ouagadougou	***Burkina Faso***	Threatened	Vulnerable
Phnom Penh	Cambodia	Vulnerable	Vulnerable
Yaoundé	Cameroon	Non-threatened	Non-threatened
Dalian	China	Vulnerable	Vulnerable
Guangzhou	***China***	Threatened	Vulnerable
Nanjing	***China***	Threatened	Vulnerable
Wuhan	***China***	Threatened	Vulnerable
Barranquilla	Colombia	Threatened	Threatened
Cali	Colombia	Vulnerable	Vulnerable
Kinshasa	Democratic Republic of Congo	Threatened	Threatened
Guayaquil	Ecuador	Vulnerable	Vulnerable
Alexandria	Egypt	Vulnerable	Vulnerable
Cairo	Egypt	Vulnerable	Vulnerable
Helsinki	Finland	Threatened	Threatened
Accra	***Ghana***	Non-threatened	Threatened
Kumasi	Ghana	Threatened	Threatened
Budapest	Hungary	Threatened	Threatened
Agra	India	Vulnerable	Vulnerable
Hubli-Dharwad	India	Threatened	Threatened
Kozhikode	India	Vulnerable	Vulnerable
Pune	India	Vulnerable	Vulnerable
Ranchi	India	Threatened	Threatened
Thiruvananthapuram	India	Threatened	Threatened

Table S-2.1. Continued.

Baghdad	Iraq	Vulnerable	Vulnerable
Dublin	***Ireland***	Threatened	Vulnerable
Mombasa	Kenya	Non-threatened	Non-threatened
Maputo	Mozambique	Vulnerable	Vulnerable
Abuja	Nigeria	Non-threatened	Non-threatened
Oslo	Norway	Threatened	Threatened
Panama City	***Panama***	Non-threatened	Threatened
Asunción	Paraguay	Non-threatened	Non-threatened
Brazzaville	Republic of Congo	Threatened	Threatened
Bucharest	Romania	Threatened	Threatened
Yekaterinburg	Russia	Threatened	Threatened
Freetown	***Sierra Leone***	Non-threatened	Threatened
Seoul	South Korea	Vulnerable	Vulnerable
Stockholm	Sweden	Threatened	Threatened
Allepo	Syria	Non-threatened	Non-threatened
Bangkok	Thailand	Vulnerable	Vulnerable
Kampala	Uganda	Non-threatened	Non-threatened
Atlanta	United States of America	Vulnerable	Vulnerable
Austin	United States of America	Vulnerable	Vulnerable
Baltimore	United States of America	Threatened	Threatened
Charlotte	***United States of America***	Threatened	Vulnerable
Fort Worth	United States of America	Threatened	Threatened
Louisville	United States of America	Threatened	Threatened
McAllen	United States of America	Vulnerable	Vulnerable
Minneapolis	United States of America	Vulnerable	Vulnerable
Nashville-Davidson	United States of America	Threatened	Threatened
New Orleans	United States of America	Non-threatened	Non-threatened

Table S-2.1. Continued.

New York	United States of America	Threatened	Threatened
Oklahoma City	United States of America	Threatened	Threatened
Philadelphia	United States of America	Vulnerable	Vulnerable
Pittsburgh	United States of America	Vulnerable	Vulnerable
Providence	United States of America	Threatened	Threatened
Raleigh	United States of America	Threatened	Threatened
Richmond	United States of America	Threatened	Threatened
Saint Louis	United States of America	Threatened	Threatened
Montevideo	Uruguay	Vulnerable	Vulnerable
Tashkent	Uzbekistan	Vulnerable	Vulnerable
Harare	Zimbabwe	Vulnerable	Vulnerable

* Bold, italicized items highlight those "cities-of-concern", who become more vulnerable during the 2010-2040 time period examined.

REFERENCES

1. Averyt K, Meldrum J, Caldwell P, Sun G, McNulty S, Huber-Lee A and Madden N 2013 Sectoral contributions to surface water stress in the coterminous United States Environ. Res. Lett. 8 035046
2. Döll P, Kaspar F and Lehner B 2003 A global hydrological model for deriving water availability indicators: model tuning and validation J. Hydrol. 270 105–34
3. Falkenmark M, Lundqvist J and Widstrand C 1989 Macro-scale water scarcity requires micro-scale approaches—aspects of vulnerability in semi-arid development Nat. Resour. Forum 13 258–67
4. FAO 2013 AQUASTAT DATABASE Water Resources Development and Management Service (Rome: FAO) accessed at www.fao.org/nr/water/aquastat/main/index.stm
5. Gleick P H 1993 Water in Crisis: A Guide to the World's Fresh Water Resources (Oxford: Oxford University Press)

6. Gober P 2010 Desert urbanization and the challenges of water sustainability Curr. Opin. Environ. Sustainability 2 144–50
7. GWSP 2008a Reliable monthly water discharge (V1.0) GWSP Digital Water Atlas Available online at (http://atlas.gwsp.org)
8. GWSP 2008b Environmental water requirements (V1.0) GWSP Digital Water Atlas Available online at (http://atlas.gwsp.org)
9. GWSP 2008c Water consumption for industry (V1.0) GWSP Digital Water Atlas Available online at (http://atlas.gwsp.org)
10. Jenerette G D and Larsen L 2006 A global perspective on changing sustainable urban water supplies Glob. Planet. Change 50 202–11
11. Lane M E, Kirshen P H and Vogel R M 1999 Indicators of impacts of global climate change on US water resources J. Water Resour. Plan. Manage.-Asce 125 194–204
12. Lehner B et al 2011 High-resolution mapping of the world's reservoirs and dams for sustainable river-flow management Frontiers Ecol. Environ. 9 494–502
13. Lehner B, Verdin K and Jarvis A 2008 New global hydrography derived from spaceborne elevation data Eos, Transactions (American Geophysical Union) 89 pp 93–4 Data available at (http://hydrosheds.cr.usgs.gov)
14. McDonald R I et al 2014 Water on an urban planet: urbanization and the reach of urban water infrastructure Glob. Environ. Change 27 96–105
15. McDonald R I, Douglas I, Revenga C, Hale R, Grimm N, Grönwall J and Fekete B 2011 Global urban growth and the geography of water availability, quality, and delivery Ambio 40 437–46
16. Miltner R J, White D and Yoder C 2004 The biotic integrity of streams in urban and suburbanizing landscapes Landscape Urban Plan. 69 87–100
17. Molden D (ed) 2007 Water for Food, Water for Life: A Comprehensive Assessment of Water Management in Agriculture (London: Earthscan IWMI) p 645
18. NID 2009 National Inventory of Dams (http://crunch.tec.army.mil/nidpublic/webpages/nid.cfm)
19. Oki T and Kanae S 2006 Global hydrological cycles and world water resources Science 313 1068–72
20. Padowski J C and Jawitz J W 2012 Water availability and vulnerability of 225 large cities in the United States Water Resour. Res. 48 W12602
21. Poff N L, Allan J D, Bain M B, Karr J R, Prestegaard K L, Richter B D, Sparks R E and Stromberg J C 1997 The natural flow regime BioScience 47 769–84
22. Richter B D et al 2013 Tapped out: how can cities secure their water future? Water Policy 15 335–63
23. Rogers P 1993 Integrated urban water resources management Nat. Resour. Forum 17 33–42
24. Schneider A, Friedl M A and Potere D 2009 A new map of global urban extent from MODIS satellite data Environ. Res. Lett. 4 044003
25. Siebert S, Henrich V, Frenken K and Burke J 2013 Global Map of Irrigation Areas (v5) (Rheinische Friedrich-Wilhelms-University, Bonn, Germany: Food and Agriculture Organization of the United Nations)
26. Smakhtin V, Revenga C and Döll P 2004 A pilot global assessment of environmental water requirements and scarcity Water Int. 29 307–17

27. United Nations 2012 Population Division: World Urbanization Prospects, the 2011 Revision (New York: Department of Economic and Social Affairs)
28. Velpuri N M and Senay G B 2013 Analysis of long-term trends (1950–2009) in precipitation, runoff and runoff coefficient in major urban watersheds in the United States Environ. Res. Lett. 8 024020
29. Venot J-P 2009 Rural dynamics and new challenges in the Indian water sector: the trajectory of the Krishna basin South India River Basin Trajectories: Societies, Environments and Development ed F Molle and P Webster (Wallingford, UK: CAB International) pp 214–37
30. Vörösmarty C J, Douglas E M, Green P A and Revenga C 2005 Geospatial indicators of emerging water stress: an application to Africa Ambio 230–6
31. Ward A D and Trimble S W 2003 Environmental Hydrology (Boca Raton, FL: CRC Press) pp 92–3
32. Webster P, Mollard E, Silva-Ochoa P and Vargas-Velazquez S 2009 From half-full to half-empty: the hydraulic mission oand water overexploitation in the Lerma-Chapala Basin, Mexico River Basin Trajectories: Societies, Environments and Development ed F Molle and P Webster (Wallingford, UK: CAB International) pp 75–98
33. Zomer R J, Bossio D A, Trabucco A, Yuanjie L, Gupta D C and Singh V P 2007 Trees and Water: Smallholder Agroforestry on Irrigated Lands in Northern India (Colombo, Sri Lanka: IMWI)
34. Zomer R J, Trabucco A, Bossio D A, van Straaten O and Verchot L V 2008 Climate change mitigation: a spatial analysis of global land suitability for clean development mechanism afforestation and reforestation Agric. Ecosyst. Environ. 126 67–80

AUTHOR NOTES

CHAPTER 1

Conflicts of Interest

The author declares no conflict of interest.

CHAPTER 2

Acknowledgments

We would like to thank all our co-authors, in particular Henrik Ernstson, Karin Ahrné and Jakob Lundberg. The original research was made possible by PhD-grants from the Department of Systems Ecology, Stockholm University, and by funds granted by The Swedish Research Council for the Environment, Agricultural Sciences and Spatial Planning (FORMAS), among them Urban Form-project and the SUPER-project (Colding, Barthel and Andersson), Urban Biodiversity – patterns and processes (Andersson), and through the Center of Excellence (FORMAS) (Gren).

CHAPTER 3

Acknowledgments

The authors are grateful for funding from the Riksbankens Jubileumsfond and the Nordic Centre of Excellence NORD-STAR.

Author Contributions

The first author conceived of the article aim, theoretical framework and general article structure. The second author defined the final article structure and case descriptions. The following authors contributed case study descriptions.

Conflicts of Interest

The authors declare no conflict of interest.

CHAPTER 4

Acknowledgments

This research was funded through a grants from the Social Sciences and Humanities Research Council to William Rees (Getting Serious About Urban Sustainability 410-2007-0473), and the Foreign Affairs and International Trade Canada (DFAIT) fellowship to Meidad Kis- singer ("Understanding Canada-Canadian Studies Program").

CHAPTER 5

Conflicts of Interest

The authors have declared that no competing interests exist.

Acknowledgments

We thank our reviewers for their valuable comments on the manuscript.

Author Contributions

Conceived and designed the experiments: MF JL DS KCS. Performed the experiments: MF JL DS KCS. Analyzed the data: MF JL DS KCS. Contributed reagents/materials/analysis tools: MF JL DS KCS. Wrote the paper: MF JL DS KCS.

CHAPTER 7

Acknowledgments

This research is financed by the German Environment Foundation (Deutsche Bundesstiftung Umwelt, DBU).

CHAPTER 8

Acknowledgments

The paper is a preliminary outcome of an Australian Research Council Linkage Project, ARC-LP0882637: Adaptation of Water Sensitive Urban Design to Climate Change, Changing Transport Patterns and Urban Form. The project is jointly funded by the Commonwealth Government of Australia, Gold Coast City Council, Queensland Transport and Main Roads and Queensland University of Technology. The authors would like to acknowledge abovementioned project partners and the research team's support and diligent work, and also would like to thank the reviewers for their comments that helped us to improve the manuscript.

CHAPTER 9

Acknowledgments

This work was developed under STAR Fellowship Assistance Agreement no. FP917275 awarded by the US Environmental Protection Agency (EPA). It has not been formally reviewed by EPA. The views expressed are solely those of the authors, and EPA does not endorse any products or commercial services mentioned.

CHAPTER 10

Acknowledgments

Co-funding of the project leading to these results by the European Commission within the 7th Framework Programme under Grant Number 213154 is kindly acknowledged.

CHAPTER 11

Acknowledgments

This research was supported by the Global Freshwater Initiative through Stanford's Woods Institute for the Environment. The authors would like to thank Rob McDonald and the City Water Map Initiative team for access to their database. This work was supported in part by the National Science Foundation under grant GEO/OAD-1342869 to Stanford University. Any opinions, findings, and conclusions or recommendations expressed in this material are those of the authors and do not necessarily reflect the views of the National Science Foundation.

INDEX

N

O

P

Q

R

U

V

W